An Introduction to Lie Groups and Lie Algebras

With roots in the nineteenth century, Lie theory has since found many and varied applications in mathematics and mathematical physics, to the point where it is now regarded as a classical branch of mathematics in its own right. This graduate text focuses on the study of semisimple Lie algebras, developing the necessary theory along the way.

The material covered ranges from basic definitions of Lie groups, to the theory of root systems, and classification of finite-dimensional representations of semisimple Lie algebras. Written in an informal style, this is a contemporary introduction to the subject which emphasizes the main concepts of the proofs and outlines the necessary technical details, allowing the material to be conveyed concisely.

Based on a lecture course given by the author at the State University of New York at Stony Brook, the book includes numerous exercises and worked examples and is ideal for graduate courses on Lie groups and Lie algebras.

CAMBRIDGE STUDIES IN ADVANCED MATHEMATICS

All the titles listed below can be obtained from good booksellers or from Cambridge University Press. For a complete series listing visit: http://www.cambridge.org/series/sSeries.asp?code=CSAM

Already published
60 M. P. Brodmann & R. Y. Sharp *Local cohomology*
61 J. D. Dixon et al. *Analytic pro-p groups*
62 R. Stanley *Enumerative combinatorics II*
63 R. M. Dudley *Uniform central limit theorems*
64 J. Jost & X. Li-Jost *Calculus of variations*
65 A. J. Berrick & M. E. Keating *An introduction to rings and modules*
66 S. Morosawa *Holomorphic dynamics*
67 A. J. Berrick & M. E. Keating *Categories and modules with K-theory in view*
68 K. Sato *Levy processes and infinitely divisible distributions*
69 H. Hida *Modular forms and Galois cohomology*
70 R. Iorio & V. Iorio *Fourier analysis and partial differential equations*
71 R. Blei *Analysis in integer and fractional dimensions*
72 F. Borceaux & G. Janelidze *Galois theories*
73 B. Bollobás *Random graphs*
74 R. M. Dudley *Real analysis and probability*
75 T. Sheil-Small *Complex polynomials*
76 C. Voisin *Hodge theory and complex algebraic geometry, I*
77 C. Voisin *Hodge theory and complex algebraic geometry, II*
78 V. Paulsen *Completely bounded maps and operator algebras*
79 F. Gesztesy & H. Holden *Soliton Equations and Their Algebro-Geometric Solutions, I*
81 S. Mukai *An Introduction to Invariants and Moduli*
82 G. Tourlakis *Lectures in Logic and Set Theory, I*
83 G. Tourlakis *Lectures in Logic and Set Theory, II*
84 R. A. Bailey *Association Schemes*
85 J. Carison, S. Müller-Stach & C. Peters *Period Mappings and Period Domains*
86 J. J. Duistermaat & J. A. C. Kolk *Multidimensional Real Analysis I*
87 J. J. Duistermaat & J. A. C. Kolk *Multidimensional Real Analysis II*
89 M. Golumbic & A. Trenk *Tolerance Graphs*
90 L. Harper *Global Methods for Combinatorial Isoperimetric Problems*
91 I. Moerdijk & J. Mrcun *Introduction to Foliations and Lie Groupoids*
92 J. Kollar, K. E. Smith & A. Corti *Rational and Nearly Rational Varieties*
93 D. Applebaum *Levy Processes and Stochastic Calculus*
94 B. Conrad *Modular Forms and the Ramanujan Conjecture*
95 M. Schechter *An Introduction to Nonlinear Analysis*
96 R. Carter *Lie Algebras of Finite and Affine Type*
97 H. L. Montgomery, R. C. Vaughan & M. Schechter *Multiplicative Number Theory I*
98 I. Chavel *Riemannian Geometry*
99 D. Goldfeld *Automorphic Forms and L-Functions for the Group GL(n,R)*
100 M. Marcus & J. Rosen *Markov Processes. Gaussian Processes, and Local Times*
101 P. Gille & T. Szamuely *Central Simple Algebras and Galois Cohomology*
102 J. Bertoin *Random Fragmentation and Coagulation Processes*
103 E. Frenkel *Langlands Correspondence for Loop Groups*
104 A. Ambrosetti & A. Malchiodi *Nonlinear Analysis and Semilinear Elliptic Problems*
105 T. Tao & V. H. Vu *Additive Combinatorics*
106 E. B. Davies *Linear Operators and their Spectra*
107 K. Kodaira *Complex Analysis*
108 T. Ceccherini-Silberstein, F. Scarabotti & F. Tolli *Harmonic Analysis on Finite Groups*
109 H. Geiges *An Introduction to Contact Topology*
110 J. Faraut *Analysis on Lie Groups*
111 E. Park *Complex Topological K-Theory*
112 D. W. Stroock *Partial Differential Equations for Probabilists*
113 A. Kirillov *An Introduction to Lie Groups and Lie Algebras*

Preface

This book is an introduction to the theory of Lie groups and Lie algebras, with emphasis on the theory of semisimple Lie algebras. It can serve as a basis for a two-semester graduate course or – omitting some material – as a basis for a rather intensive one-semester course. The book includes a large number of exercises.

The material covered in the book ranges from basic definitions of Lie groups to the theory of root systems and highest weight representations of semisimple Lie algebras; however, to keep book size small, the structure theory of semisimple and compact Lie groups is not covered.

Exposition follows the style of famous Serre's textbook on Lie algebras [47]: we tried to make the book more readable by stressing ideas of the proofs rather than technical details. In many cases, details of the proofs are given in exercises (always providing sufficient hints so that good students should have no difficulty completing the proof). In some cases, technical proofs are omitted altogether; for example, we do not give proofs of Engel's or Poincare–Birkhoff–Witt theorems, instead providing an outline of the proof. Of course, in such cases we give references to books containing full proofs.

It is assumed that the reader is familiar with basics of topology and differential geometry (manifolds, vector fields, differential forms, fundamental groups, covering spaces) and basic algebra (rings, modules). Some parts of the book require knowledge of basic homological algebra (short and long exact sequences, Ext spaces).

Errata for this book are available on the book web page at
http://www.math.sunysb.edu/~kirillov/liegroups/.

1

Introduction

In any algebra textbook, the study of group theory is usually mostly concerned with the theory of finite, or at least finitely generated, groups. This is understandable: such groups are much easier to describe. However, most groups which appear as groups of symmetries of various geometric objects are not finite: for example, the group $SO(3, \mathbb{R})$ of all rotations of three-dimensional space is not finite and is not even finitely generated. Thus, much of material learned in basic algebra course does not apply here; for example, it is not clear whether, say, the set of all morphisms between such groups can be explicitly described.

The theory of Lie groups answers these questions by replacing the notion of a finitely generated group by that of a Lie group – a group which at the same time is a finite-dimensional manifold. It turns out that in many ways such groups can be described and studied as easily as finitely generated groups – or even easier. The key role is played by the notion of a Lie algebra, the tangent space to G at identity. It turns out that the group operation on G defines a certain bilinear skew-symmetric operation on $\mathfrak{g} = T_1G$; axiomatizing the properties of this operation gives a definition of a Lie algebra.

The fundamental result of the theory of Lie groups is that many properties of Lie groups are completely determined by the properties of corresponding Lie algebras. For example, the set of morphisms between two (connected and simply connected) Lie groups is the same as the set of morphisms between the corresponding Lie algebras; thus, describing them is essentially reduced to a linear algebra problem.

Similarly, Lie algebras also provide a key to the study of the structure of Lie groups and their representations. In particular, this allows one to get a complete classification of a large class of Lie groups (semisimple and more generally, reductive Lie groups; this includes all compact Lie groups and all classical Lie groups such as $SO(n, \mathbb{R})$) in terms of relatively simple geometric objects, so-called root systems. This result is considered by many mathematicians (including the author of this book) to be one of the most beautiful achievements in all of mathematics. We will cover it in Chapter 7.

To conclude this introduction, we will give a simple example which shows how Lie groups naturally appear as groups of symmetries of various objects – and how one can use the theory of Lie groups and Lie algebras to make use of these symmetries.

Let $S^2 \subset \mathbb{R}^3$ be the unit sphere. Define the Laplace operator Δ_{sph} : $C^\infty(S^2) \to C^\infty(S^2)$ by $\Delta_{\mathrm{sph}} f = (\Delta \tilde{f})|_{S^2}$, where $\tilde{f}$ is the result of extending f to $\mathbb{R}^3 - \{0\}$ (constant along each ray), and Δ is the usual Laplace operator in $\mathbb{R}^3$. It is easy to see that Δ_{sph} is a second-order differential operator on the sphere; one can write explicit formulas for it in the spherical coordinates, but they are not particularly nice.

For many applications, it is important to know the eigenvalues and eigenfunctions of Δ_{sph}. In particular, this problem arises in quantum mechanics: the eigenvalues are related to the energy levels of a hydrogen atom in quantum mechanical description. Unfortunately, trying to find the eigenfunctions by brute force gives a second-order differential equation which is very difficult to solve.

However, it is easy to notice that this problem has some symmetry – namely, the group $\mathrm{SO}(3, \mathbb{R})$ acting on the sphere by rotations. How can one use this symmetry?

If we had just one symmetry, given by some rotation $R: S^2 \to S^2$, we could consider its action on the space of complex-valued functions $C^\infty(S^2, \mathbb{C})$. If we could diagonalize this operator, this would help us study Δ_{sph}: it is a general result of linear algebra that if A, B are two commuting operators, and A is diagonalizable, then B must preserve eigenspaces for A. Applying this to pair R, Δ_{sph}, we get that Δ_{sph} preserves eigenspaces for R, so we can diagonalize Δ_{sph} independently in each of the eigenspaces.

However, this will not solve the problem: for each individual rotation R, the eigenspaces will still be too large (in fact, infinite-dimensional), so diagonalizing Δ_{sph} in each of them is not very easy either. This is not surprising: after all, we only used one of many symmetries. Can we use all of rotations $R \in \mathrm{SO}(3, \mathbb{R})$ simultaneously?

This, however, presents two problems.

- $\mathrm{SO}(3, \mathbb{R})$ is not a finitely generated group, so apparently we will need to use infinitely (in fact uncountably) many different symmetries and diagonalize each of them.
- $\mathrm{SO}(3, \mathbb{R})$ is not commutative, so different operators from $\mathrm{SO}(3, \mathbb{R})$ can not be diagonalized simultaneously.

The goal of the theory of Lie groups is to give tools to deal with these (and similar) problems. In short, the answer to the first problem is that $SO(3, \mathbb{R})$ is in a certain sense finitely generated – namely, it is generated by three generators, "infinitesimal rotations" around x, y, z axes (see details in Example 3.10).

The answer to the second problem is that instead of decomposing the $C^\infty(S^2, \mathbb{C})$ into a direct sum of common eigenspaces for operators $R \in SO(3, \mathbb{R})$, we need to decompose it into "irreducible representations" of $SO(3, \mathbb{R})$. In order to do this, we need to develop the theory of representations of $SO(3, \mathbb{R})$. We will do this and complete the analysis of this example in Section 4.8.

2

Lie groups: basic definitions

2.1. Reminders from differential geometry

This book assumes that the reader is familiar with basic notions of differential geometry, as covered for example, in [49]. For reader's convenience, in this section we briefly remind some definitions and fix notation.

Unless otherwise specified, all manifolds considered in this book will be C^∞ real manifolds; the word "smooth" will mean C^∞. All manifolds we will consider will have at most countably many connected components.

For a manifold M and a point $m \in M$, we denote by $T_m M$ the tangent space to M at point m, and by TM the tangent bundle to M. The space of vector fields on M (i.e., global sections of TM) is denoted by $\text{Vect}(M)$. For a morphism $f : X \to Y$ and a point $x \in X$, we denote by $f_* : T_x X \to T_{f(x)} Y$ the corresponding map of tangent spaces.

Recall that a morphism $f : X \to Y$ is called an *immersion* if $\text{rank} f_* = \dim X$ for every point $x \in X$; in this case, one can choose local coordinates in a neighborhood of $x \in X$ and in a neighborhood of $f(x) \in Y$ such that f is given by $f(x_1, \dots x_n) = (x_1, \dots, x_n, 0, \dots 0)$.

An *immersed submanifold* in a manifold M is a subset $N \subset M$ with a structure of a manifold (not necessarily the one inherited from M!) such that inclusion map $i : N \hookrightarrow M$ is an immersion. Note that the manifold structure on N is part of the data: in general, it is not unique. However, it is usually suppressed in the notation. Note also that for any point $p \in N$, the tangent space to N is naturally a subspace of tangent space to $M : T_p N \subset T_p M$.

An *embedded submanifold* $N \subset M$ is an immersed submanifold such that the inclusion map $i : N \hookrightarrow M$ is a homeomorphism. In this case the smooth structure on N is uniquely determined by the smooth structure on M.

Following Spivak, we will use the word "submanifold" for *embedded* submanifolds (note that many books use word submanifold for immersed submanifolds).

All of the notions above have complex analogs, in which real manifolds are replaced by complex analytic manifolds and smooth maps by holomorphic maps. We refer the reader to [49] for details.

2.2. Lie groups, subgroups, and cosets

Definition 2.1. A (real) Lie group is a set G with two structures: G is a group and G is a manifold. These structures agree in the following sense: multiplication map $G \times G \to G$ and inversion map $G \to G$ are smooth maps.

A morphism of Lie groups is a smooth map which also preserves the group operation: $f(gh) = f(g)f(h), f(1) = 1$. We will use the standard notation $\operatorname{Im} f$, $\operatorname{Ker} f$ for image and kernel of a morphism.

The word "real" is used to distinguish these Lie groups from complex Lie groups defined below. However, it is frequently omitted: unless one wants to stress the difference with complex case, it is common to refer to real Lie groups as simply Lie groups.

Remark 2.2. One can also consider other classes of manifolds: C^1, C^2, analytic. It turns out that all of them are equivalent: every C^0 Lie group has a unique analytic structure. This is a highly non-trivial result (it was one of Hilbert's 20 problems), and we are not going to prove it (the proof can be found in the book [39]). Proof of a weaker result, that C^2 implies analyticity, is much easier and can be found in [10, Section 1.6]. In this book, "smooth" will be always understood as C^∞.

In a similar way, one defines complex Lie groups.

Definition 2.3. A complex Lie group is a set G with two structures: G is a group and G is a complex analytic manifold. These structures agree in the following sense: multiplication map $G \times G \to G$ and inversion map $G \to G$ are analytic maps.

A morphism of complex Lie groups is an analytic map which also preserves the group operation: $f(gh) = f(g)f(h), f(1) = 1$.

Remark 2.4. Throughout this book, we try to treat both real and complex cases simultaneously. Thus, most theorems in this book apply both to real and complex Lie groups. In such cases, we will say "let G be real or complex Lie group ..." or "let G be a Lie group over $\mathbb{K}$ ...", where $\mathbb{K}$ is the base field: $\mathbb{K} = \mathbb{R}$ for real Lie groups and $\mathbb{K} = \mathbb{C}$ for complex Lie groups.

When talking about complex Lie groups, "submanifold" will mean "complex analytic submanifold", tangent spaces will be considered as complex vector spaces, all morphisms between manifolds will be assumed holomorphic, etc.

Example 2.5. The following are examples of Lie groups:

(1) $\mathbb{R}^n$, with the group operation given by addition
(2) $\mathbb{R}^* = \mathbb{R} \setminus \{0\}$, $\times$
     $\mathbb{R}_+ = \{x \in \mathbb{R} | x > 0\}$, $\times$
(3) $S^1 = \{z \in \mathbb{C} : |z| = 1\}$, $\times$
(4) $\mathrm{GL}(n, \mathbb{R}) \subset \mathbb{R}^{n^2}$. Many of the groups we will consider will be subgroups of $\mathrm{GL}(n, \mathbb{R})$ or $\mathrm{GL}(n, \mathbb{C})$.
(5) $\mathrm{SU}(2) = \{A \in \mathrm{GL}(2, \mathbb{C}) \mid A\bar{A}^t = 1, \det A = 1\}$. Indeed, one can easily see that

$$\mathrm{SU}(2) = \left\{ \begin{pmatrix} \alpha & \beta \\ -\bar{\beta} & \bar{\alpha} \end{pmatrix} : \alpha, \beta \in \mathbb{C}, |\alpha|^2 + |\beta|^2 = 1 \right\}.$$

Writing $\alpha = x_1 + i x_2, \beta = x_3 + i x_4, x_i \in \mathbb{R}$, we see that $\mathrm{SU}(2)$ is diffeomorphic to $S^3 = \{x_1^2 + \cdots + x_4^2 = 1\} \subset \mathbb{R}^4$.
(6) In fact, all usual groups of linear algebra, such as $\mathrm{GL}(n, \mathbb{R})$, $\mathrm{SL}(n, \mathbb{R})$, $\mathrm{O}(n, \mathbb{R})$, $\mathrm{U}(n)$, $\mathrm{SO}(n, \mathbb{R})$, $\mathrm{SU}(n)$, $\mathrm{Sp}(n, \mathbb{R})$ are (real or complex) Lie groups. This will be proved later (see Section 2.7).

Note that the definition of a Lie group does not require that G be connected. Thus, any finite group is a 0-dimensional Lie group. Since the theory of finite groups is complicated enough, it makes sense to separate the finite (or, more generally, discrete) part. It can be done as follows.

Theorem 2.6. *Let G be a real or complex Lie group. Denote by G^0 the connected component of identity. Then G^0 is a normal subgroup of G and is a Lie group itself (real or complex, respectively). The quotient group G/G^0 is discrete.*

Proof. We need to show that G^0 is closed under the operations of multiplication and inversion. Since the image of a connected topological space under a continuous map is connected, the inversion map i must take G^0 to one component of G, that which contains $i(1) = 1$, namely G^0. In a similar way one shows that G^0 is closed under multiplication.

To check that this is a normal subgroup, we must show that if $g \in G$ and $h \in G^0$, then $ghg^{-1} \in G^0$. Conjugation by g is continuous and thus will take G^0 to some connected component of G; since it fixes 1, this component is G^0.

The fact that the quotient is discrete is obvious. $\qquad\square$

This theorem mostly reduces the study of arbitrary Lie groups to the study of finite groups and connected Lie groups. In fact, one can go further and reduce the study of connected Lie groups to connected simply-connected Lie groups.

Theorem 2.7. *If G is a connected Lie group (real or complex), then its universal cover $\tilde{G}$ has a canonical structure of a Lie group (real or complex, respectively) such that the covering map $p\colon \tilde{G} \to G$ is a morphism of Lie groups whose kernel is isomorphic to the fundamental group of G: $\mathrm{Ker}\, p = \pi_1(G)$ as a group. Moreover, in this case $\mathrm{Ker}\, p$ is a discrete central subgroup in $\tilde{G}$.*

Proof. The proof follows from the following general result of topology: if M, N are connected manifolds (or, more generally, nice enough topological spaces), then any continuous map $f\colon M \to N$ can be lifted to a map of universal covers $\tilde{f}\colon \tilde{M} \to \tilde{N}$. Moreover, if we choose $m \in M, n \in N$ such that $f(m) = n$ and choose liftings $\tilde{m} \in \tilde{M}, \tilde{n} \in \tilde{N}$ such that $p(\tilde{m}) = m, p(\tilde{n}) = n$, then there is a unique lifting $\tilde{f}$ of f such that $\tilde{f}(\tilde{m}) = \tilde{n}$.

Now let us choose some element $\tilde{1} \in \tilde{G}$ such that $p(\tilde{1}) = 1 \in G$. Then, by the above theorem, there is a unique map $\tilde{i}\colon \tilde{G} \to \tilde{G}$ which lifts the inversion map $i\colon G \to G$ and satisfies $\tilde{i}(\tilde{1}) = \tilde{1}$. In a similar way one constructs the multiplication map $\tilde{G} \times \tilde{G} \to \tilde{G}$. Details are left to the reader.

Finally, the fact that $\mathrm{Ker}\, p$ is central follows from results of Exercise 2.2. $\square$

Definition 2.8. A closed Lie subgroup H of a (real or complex) Lie group G is a subgroup which is also a submanifold (for complex Lie groups, it is must be a complex submanifold).

Note that the definition does not require that H be a closed subset in G; thus, the word "closed" requires some justification which is given by the following result.

Theorem 2.9.

(1) *Any closed Lie subgroup is closed in G.*
(2) *Any closed subgroup of a Lie group is a closed real Lie subgroup.*

Proof. The proof of the first part is given in Exercise 2.1. The second part is much harder and will not be proved here (and will not be used in this book). The proof uses the technique of Lie algebras and can be found, for example, in [10, Corollary 1.10.7]. We will give a proof of a weaker but sufficient for our purposes result later (see Section 3.6). $\square$

Corollary 2.10.

(1) *If G is a connected Lie group (real or complex) and U is a neighborhood of 1, then U generates G.*
(2) *Let $f: G_1 \to G_2$ be a morphism of Lie groups (real or complex), with G_2 connected, such that $f_*: T_1 G_1 \to T_1 G_2$ is surjective. Then f is surjective.*

Proof. (1) Let H be the subgroup generated by U. Then H is open in G: for any element $h \in H$, the set $h \cdot U$ is a neighborhood of h in G. Since it is an open subset of a manifold, it is a submanifold, so H is a closed Lie subgroup. Therefore, by Theorem 2.9 it is closed, and is nonempty, so $H = G$.

(2) Given the assumption, the inverse function theorem says that f is surjective onto some neighborhood U of $1 \in G_2$. Since an image of a group morphism is a subgroup, and U generates G_2, f is surjective. $\qquad\square$

As in the theory of discrete groups, given a closed Lie subgroup $H \subset G$, we can define the notion of cosets and define the coset space G/H as the set of equivalence classes. The following theorem shows that the coset space is actually a manifold.

Theorem 2.11.

(1) *Let G be a (real or complex) Lie group of dimension n and $H \subset G$ a closed Lie subgroup of dimension k. Then the coset space G/H has a natural structure of a manifold of dimension $n - k$ such that the canonical map $p: G \to G/H$ is a fiber bundle, with fiber diffeomorphic to H. The tangent space at $\bar{1} = p(1)$ is given by $T_{\bar{1}}(G/H) = T_1 G / T_1 H$.*
(2) *If H is a normal closed Lie subgroup then G/H has a canonical structure of a Lie group (real or complex, respectively).*

Proof. Denote by $p: G \to G/H$ the canonical map. Let $g \in G$ and $\bar{g} = p(g) \in G/H$. Then the set $g \cdot H$ is a submanifold in G as it is an image of H under diffeomorphism $x \mapsto gx$. Choose a submanifold $M \subset G$ such that $g \in M$ and M is transversal to the manifold gH, i.e. $T_g G = (T_g(gH)) \oplus T_g M$ (this implies that $\dim M = \dim G - \dim H$). Let $U \subset M$ be a sufficiently small neighborhood of g in M. Then the set $UH = \{uh \mid u \in U, h \in H\}$ is open in G (which easily follows from inverse function theorem applied to the map $U \times H \to G$). Consider $\bar{U} = p(U)$; since $p^{-1}(\bar{U}) = UH$ is open, $\bar{U}$ is an open neighborhood of $\bar{g}$ in G/H and the map $U \to \bar{U}$ is a homeomorphism. This gives a local chart for G/H and at the same time shows that $G \to G/H$ is

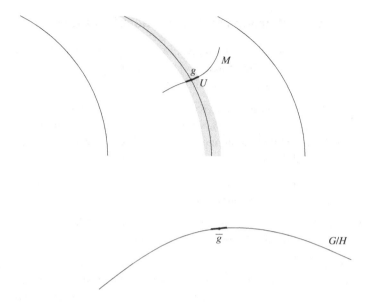

Figure 2.1 Fiber bundle $G \to G/H$

a fiber bundle with fiber H (Figure 2.1). We leave it to the reader to show that transition functions between such charts are smooth (respectively, analytic) and that the smooth structure does not depend on the choice of g, M.

This argument also shows that the kernel of the projection $p_* \colon T_g G \to T_{\bar{g}}(G/H)$ is equal to $T_g(gH)$. In particular, for $g = 1$ this gives an isomorphism $T_{\bar{1}}(G/H) = T_1 G / T_1 H$. $\qquad\qquad\square$

Corollary 2.12. *Let H be a closed Lie subgroup of a Lie group G.*

(1) *If H is connected, then the set of connected components $\pi_0(G) = \pi_0(G/H)$. In particular, if $H, G/H$ are connected, then so is G.*
(2) *If G, H are connected, then there is an exact sequence of fundamental groups*

$$\pi_2(G/H) \to \pi_1(H) \to \pi_1(G) \to \pi_1(G/H) \to \{1\}.$$

This corollary follows from more general long exact sequence of homotopy groups associated with any fiber bundle (see [17, Section 4.2]). We will later use it to compute fundamental groups of classical groups such as $\mathrm{GL}(n, \mathbb{K})$.

2.3. Lie subgroups and homomorphism theorem

For many purposes, the notion of closed Lie subgroup introduced above is too restrictive. For example, the image of a morphism may not be a closed Lie subgroup, as the following example shows.

Example 2.13. Let $G_1 = \mathbb{R}, G_2 = T^2 = \mathbb{R}^2/\mathbb{Z}^2$. Define the map $f: G_1 \to G_2$ by $f(t) = (t \mod \mathbb{Z}, \alpha t \mod \mathbb{Z})$, where α is some fixed irrational number. Then it is well-known that the image of this map is everywhere dense in T^2 (it is sometimes called the *irrational winding* on the torus).

Thus, it is useful to introduce a more general notion of a subgroup. Recall the definition of immersed submanifold (see Section 2.1).

Definition 2.14. An *Lie subgroup* in a (real or complex) Lie group $H \subset G$ is an immersed submanifold which is also a subgroup.

It is easy to see that in such a situation H is itself a Lie group (real or complex, respectively) and the inclusion map $i: H \hookrightarrow G$ is a morphism of Lie groups.

Clearly, every closed Lie subgroup is a Lie subgroup, but the converse is not true: the image of the map $\mathbb{R} \to T^2$ constructed in Example 2.13 is a Lie subgroup which is not closed. It can be shown if a Lie subgroup is closed in G, then it is automatically a closed Lie subgroup in the sense of Definition 2.8, which justifies the name. We do not give a proof of this statement as we are not going to use it.

With this new notion of a subgroup we can formulate an analog of the standard homomorphism theorems.

Theorem 2.15. *Let $f: G_1 \to G_2$ be a morphism of (real or complex) Lie groups. Then $H = \text{Ker} f$ is a normal closed Lie subgroup in G_1, and f gives rise to an injective morphism $G_1/H \to G_2$, which is an immersion; thus, $\text{Im} f$ is a Lie subgroup in G_2. If $\text{Im} f$ is an (embedded) submanifold, then it is a closed Lie subgroup in G_2 and f gives an isomorphism of Lie groups $G_1/H \simeq \text{Im} f$.*

The easiest way to prove this theorem is by using the theory of Lie algebras which we will develop in the next chapter; thus, we postpone the proof until the next chapter (see Corollary 3.30).

2.4. Action of Lie groups on manifolds and representations

The primary reason why Lie groups are so frequently used is that they usually appear as symmetry groups of various geometric objects. In this section, we will show several examples.

Definition 2.16. An action of a real Lie group G on a manifold M is an assignment to each $g \in G$ a diffeomorphism $\rho(g) \in \mathrm{Diff}\, M$ such that $\rho(1) = \mathrm{id}, \rho(gh) = \rho(g)\rho(h)$ and such that the map

$$G \times M \to M : (g, m) \mapsto \rho(g).m$$

is a smooth map.

A holomorphic action of a complex Lie group G on a complex manifold M is an assignment to each $g \in G$ an invertible holomorphic map $\rho(g) \in \mathrm{Diff}\, M$ such that $\rho(1) = \mathrm{id}, \rho(gh) = \rho(g)\rho(h)$ and such that the map

$$G \times M \to M : (g, m) \mapsto \rho(g).m$$

is holomorphic.

Example 2.17.

(1) The group $\mathrm{GL}(n, \mathbb{R})$ (and thus, any its closed Lie subgroup) acts on $\mathbb{R}^n$.
(2) The group $\mathrm{O}(n, \mathbb{R})$ acts on the sphere $S^{n-1} \subset \mathbb{R}^n$. The group $\mathrm{U}(n)$ acts on the sphere $S^{2n-1} \subset \mathbb{C}^n$.

Closely related with the notion of a group action on a manifold is the notion of a representation.

Definition 2.18. A representation of a (real or complex) Lie group G is a vector space V (complex if G is complex, and either real or complex if G is real) together with a group morphism $\rho : G \to GL(V)$. If V is finite-dimensional, we require that ρ be smooth (respectively, analytic), so it is a morphism of Lie groups. A morphism between two representations V, W of the same group G is a linear map $f : V \to W$ which commutes with the group action: $f \rho_V(g) = \rho_W(g)f$.

In other words, we assign to every $g \in G$ a linear map $\rho(g) : V \to V$ so that $\rho(g)\rho(h) = \rho(gh)$.

We will frequently use the shorter notation $g.m, g.v$ instead of $\rho(g).m$ in the cases when there is no ambiguity about the representation being used.

Remark 2.19. Note that we frequently consider representations on a *complex* vector space V, even for a real Lie group G.

Any action of the group G on a manifold M gives rise to several representations of G on various vector spaces associated with M:

(1) Representation of G on the (infinite-dimensional) space of functions $C^\infty(M)$ (in real case) or the space of holomorphic functions $\mathcal{O}(M)$ (in

complex case) defined by

$$(\rho(g)f)(m) = f(g^{-1}.m) \tag{2.1}$$

(note that we need g^{-1} rather than g to satisfy $\rho(g)\rho(h) = \rho(gh)$).

(2) Representation of G on the (infinite-dimensional) space of vector fields $\text{Vect}(M)$ defined by

$$(\rho(g).v)(m) = g_*(v(g^{-1}.m)). \tag{2.2}$$

In a similar way, we define the action of G on the spaces of differential forms and other types of tensor fields on M.

(3) Assume that $m \in M$ is a fixed point: $g.m = m$ for any $g \in G$. Then we have a canonical action of G on the tangent space T_mM given by $\rho(g) = g_* \colon T_mM \to T_mM$, and similarly for the spaces T_m^*M, $\bigwedge^k T_m^*M$.

2.5. Orbits and homogeneous spaces

Let G be a Lie group acting on a manifold M (respectively, a complex Lie group acting on a complex manifold M). Then for every point $m \in M$ we define its *orbit* by $\mathcal{O}m = Gm = \{g.m \mid g \in G\}$ and stabilizer by

$$G_m = \{g \in G \mid g.m = m\}. \tag{2.3}$$

Theorem 2.20. *Let M be a manifold with an action of a Lie group G (respectively, a complex manifold with an action of complex Lie group G). Then for any $m \in M$ the stabilizer G_m is a closed Lie subgroup in G, and $g \mapsto g.m$ is an injective immersion $G/G_m \hookrightarrow M$ whose image coincides with the orbit $\mathcal{O}_m$.*

Proof. The fact that the orbit is in bijection with G/G_m is obvious. For the proof of the fact that G_m is a closed Lie subgroup, we could just refer to Theorem 2.9. However, this would not help proving that $G/G_m \to M$ is an immersion. Both of these statements are easiest proved using the technique of Lie algebras; thus, we postpone the proof until later (see Theorem 3.29). □

Corollary 2.21. *The orbit $\mathcal{O}_m$ is an immersed submanifold in M, with tangent space $T_m\mathcal{O}_m = T_1G/T_1G_m$. If $\mathcal{O}_m$ is a submanifold, then $g \mapsto g.m$ is a diffeomorphism $G/G_m \xrightarrow{\sim} \mathcal{O}_m$.*

An important special case is when the action of G is transitive, i.e. when there is only one orbit.

Definition 2.22. A G-homogeneous space is a manifold with a transitive action of G.

As an immediate corollary of Corollary 2.21, we see that each homogeneous space is diffeomorphic to a coset space G/H. Combining it with Theorem 2.11, we get the following result.

Corollary 2.23. *Let M be a G-homogeneous space and choose $m \in M$. Then the map $G \to M : g \mapsto gm$ is a fiber bundle over M with fiber G_m.*

Example 2.24.

(1) Consider the action of $SO(n, \mathbb{R})$ on the sphere $S^{n-1} \subset \mathbb{R}^n$. Then it is a homogeneous space, so we have a fiber bundle

$$SO(n-1, \mathbb{R}) \longrightarrow SO(n, \mathbb{R})$$
$$\downarrow$$
$$S^{n-1}$$

(2) Consider the action of $SU(n)$ on the sphere $S^{2n-1} \subset \mathbb{C}^n$. Then it is a homogeneous space, so we have a fiber bundle

$$SU(n-1) \longrightarrow SU(n)$$
$$\downarrow$$
$$S^{2n-1}$$

In fact, the action of G can be used to define smooth structure on a set. Indeed, if M is a set (no smooth structure yet) with a transitive action of a Lie group G, then M is in bijection with G/H, $H = \text{Stab}_G(m)$ and thus, by Theorem 2.11, M has a canonical structure of a manifold of dimension equal to $\dim G - \dim H$.

Example 2.25. Define a *flag* in $\mathbb{R}^n$ to be a sequence of subspaces

$$\{0\} \subset V_1 \subset V_2 \subset \cdots \subset V_n = \mathbb{R}^n, \qquad \dim V_i = i.$$

Let $\mathcal{F}_n(\mathbb{R})$ be the set of all flags in $\mathbb{R}^n$. It turns out that $\mathcal{F}_n(\mathbb{R})$ has a canonical structure of a smooth manifold, which is called the *flag manifold* (or sometimes *flag variety*). The easiest way to define it is to note that we have an obvious action of the group $GL(n, \mathbb{R})$ on $\mathcal{F}_n(\mathbb{R})$. This action is transitive: by a change of basis, any flag can be identified with the standard flag

$$V^{st} = \big(\{0\} \subset \langle e_1 \rangle \subset \langle e_1, e_2 \rangle \subset \cdots \subset \langle e_1, \ldots, e_{n-1} \rangle \subset \mathbb{R}^n\big),$$

where $\langle e_1, \ldots, e_k \rangle$ stands for the subspace spanned by $e_1, \ldots, e_k$. Thus, $\mathcal{F}_n(\mathbb{R})$ can be identified with the coset space $GL(n, \mathbb{R})/B(n, \mathbb{R})$, where $B(n, \mathbb{R}) =$ Stab V^{st} is the group of all invertible upper-triangular matrices. Therefore, $\mathcal{F}_n$ is a manifold of dimension equal to $n^2 - (n(n+1))/2 = n(n-1)/2$.

Finally, we should say a few words about taking the quotient by the action of a group. In many cases when we have an action of a group G on a manifold M one would like to consider the quotient space, i.e. the set of all G-orbits. This set is commonly denoted by M/G. It has a canonical quotient space topology. However, this space can be very singular, even if G is a Lie group; for example, it can be non-Hausdorff. For example, for the group $G = GL(n, \mathbb{C})$ acting on the set of all $n \times n$ matrices by conjugation the set of orbits is described by Jordan canonical form. However, it is well-known that by a small perturbation, any matrix can be made diagonalizable. Thus, if X is a diagonalizable matrix and Y is a non-diagonalizable matrix with the same eigenvalues as X, then any neighborhood of the orbit of Y contains points from orbit of X.

There are several ways of dealing with this problem. One of them is to impose additional requirements on the action, for example assuming that the action is proper. In this case it can be shown that M/G is indeed a Hausdorff topological space, and under some additional conditions, it is actually a manifold (see [10, Section 2]). Another approach, usually called Geometric Invariant Theory, is based on using the methods of algebraic geometry (see [40]). Both of these methods go beyond the scope of this book.

2.6. Left, right, and adjoint action

Important examples of group action are the following actions of G on itself:

Left action: $L_g : G \to G$ is defined by $L_g(h) = gh$
Right action: $R_g : G \to G$ is defined by $R_g(h) = hg^{-1}$
Adjoint action: $\text{Ad } g : G \to G$ is defined by $\text{Ad } g(h) = ghg^{-1}$.

One easily sees that left and right actions are transitive; in fact, each of them is simply transitive. It is also easy to see that the left and right actions commute and that $\text{Ad } g = L_g R_g$.

As mentioned in Section 2.4, each of these actions also defines the action of G on the spaces of functions, vector fields, forms, etc. on G. For simplicity, for a tangent vector $v \in T_m G$, we will frequently write just $g.v \in T_{gm}G$ instead of the technically more accurate but cumbersome notation $(L_g)_* v$. Similarly, we will write $v.g$ for $(R_{g^{-1}})_* v$. This is justified by Exercise 2.6, where it is shown that for matrix groups this notation agrees with the usual multiplication of matrices.

Since the adjoint action preserves the identity element $1 \in G$, it also defines an action of G on the (finite-dimensional) space $T_1 G$. Slightly abusing the notation, we will denote this action also by

$$\text{Ad } g : T_1 G \to T_1 G. \tag{2.4}$$

Definition 2.26. A vector field $v \in \text{Vect}(G)$ is *left-invariant* if $g.v = v$ for every $g \in G$, and right-invariant if $v.g = v$ for every $g \in G$. A vector field is called bi-invariant if it is both left- and right-invariant.

In a similar way one defines left- , right-, and bi-invariant differential forms and other tensors.

Theorem 2.27. *The map $v \mapsto v(1)$ (where 1 is the identity element of the group) defines an isomorphism of the vector space of left-invariant vector fields on G with the vector space $T_1 G$, and similarly for right-invariant vector fields.*

Proof. It suffices to prove that every $x \in T_1 G$ can be uniquely extended to a left-invariant vector field on G. Let us define the extension by $v(g) = g.x \in T_g G$. Then one easily sees that the so-defined vector field is left-invariant, and $v(1) = x$. This proves the existence of an extension; uniqueness is obvious. $\square$

Describing bi-invariant vector fields on G is more complicated: any $x \in T_1 G$ can be uniquely extended to a left-invariant vector field and to a right-invariant vector field, but these extensions may differ.

Theorem 2.28. *The map $v \mapsto v(1)$ defines an isomorphism of the vector space of bi-invariant vector fields on G with the vector space of invariants of adjoint action:*

$$(T_1 G)^{\text{Ad } G} = \{ x \in T_1 G \mid \text{Ad } g(x) = x \text{ for all } g \in G \}.$$

The proof of this result is left to the reader. Note also that a similar result holds for other types of tensor fields: covector fields, differential forms, etc.

2.7. Classical groups

In this section, we discuss the so-called classical groups, or various sub-groups of the general linear group which are frequently used in linear algebra. Traditionally, the name "classical groups" is applied to the following groups:

- $GL(n, \mathbb{K})$ (here and below, $\mathbb{K}$ is either $\mathbb{R}$, which gives a real Lie group, or $\mathbb{C}$, which gives a complex Lie group)
- $SL(n, \mathbb{K})$
- $O(n, \mathbb{K})$
- $SO(n, \mathbb{K})$ and more general groups $SO(p, q; \mathbb{R})$.
- $Sp(n, \mathbb{K}) = \{A : \mathbb{K}^{2n} \to \mathbb{K}^{2n} \mid \omega(Ax, Ay) = \omega(x, y)\}$. Here $\omega(x, y)$ is the skew-symmetric bilinear form $\sum_{i=1}^{n} x_i y_{i+n} - y_i x_{i+n}$ (which, up to a change of basis, is the unique nondegenerate skew-symmetric bilinear form on $\mathbb{K}^{2n}$). Equivalently, one can write $\omega(x, y) = (Jx, y)$, where $(,)$ is the standard symmetric bilinear form on $\mathbb{K}^{2n}$ and

$$J = \begin{pmatrix} 0 & -I_n \\ I_n & 0 \end{pmatrix}. \tag{2.5}$$

  Note that there is some ambiguity with the notation for symplectic group: the group we denoted $Sp(n, \mathbb{K})$ would be written in some books as $Sp(2n, \mathbb{K})$.
- $U(n)$ (note that this is a real Lie group, even though its elements are matrices with complex entries)
- $SU(n)$
- Group of unitary quaternionic transformations $Sp(n) = Sp(n, \mathbb{C}) \cap SU(2n)$. Another description of this group, which explains its relation with quaternions, is given in Exercise 2.15.

  This group is a "compact form" of the group $Sp(n, \mathbb{C})$ in the sense we will describe later (see Exercise 3.16).

We have already shown that $GL(n)$ and $SU(2)$ are Lie groups. In this section, we will show that each of the classical groups listed above is a Lie group and will find their dimensions.

A straightforward approach, based on the implicit function theorem, is hard: for example, $SO(n, \mathbb{K})$ is defined by $n^2 + 1$ equations in $\mathbb{K}^{n^2}$, and finding the rank of this system is not an easy task. We could just refer to the theorem about closed subgroups; this would prove that each of them is a Lie group, but would give us no other information – not even the dimension of G. Thus, we will need another approach.

Our approach is based on the use of exponential map. Recall that for matrices, the exponential map is defined by

$$\exp(x) = \sum_0^\infty \frac{x^k}{k!}. \tag{2.6}$$

It is well-known that this power series converges and defines an analytic map $\mathfrak{gl}(n, \mathbb{K}) \to \mathfrak{gl}(n, \mathbb{K})$, where $\mathfrak{gl}(n, \mathbb{K})$ is the set of all $n \times n$ matrices. In a similar way, we define the logarithmic map by

$$\log(1 + x) = \sum_1^\infty \frac{(-1)^{k+1} x^k}{k}. \tag{2.7}$$

So defined, log is an analytic map defined in a neighborhood of $1 \in \mathfrak{gl}(n, \mathbb{K})$.

The following theorem summarizes the properties of exponential and logarithmic maps. Most of the properties are the same as for numbers; however, there are also some differences due to the fact that multiplication of matrices is not commutative. All of the statements of this theorem apply equally in real and complex cases.

Theorem 2.29.

(1) $\log(\exp(x)) = x$; $\exp(\log(X)) = X$ *whenever they are defined.*
(2) $\exp(x) = 1 + x + \dots$ *This means* $\exp(0) = 1$ *and* $d \exp(0) = \mathrm{id}$.
(3) *If* $xy = yx$ *then* $\exp(x + y) = \exp(x) \exp(y)$. *If* $XY = YX$ *then* $\log(XY) = \log(X) + \log(Y)$ *in some neighborhood of the identity. In particular, for any* $x \in \mathfrak{gl}(n, \mathbb{K})$, $\exp(x) \exp(-x) = 1$, *so* $\exp x \in \mathrm{GL}(n, \mathbb{K})$.
(4) *For fixed* $x \in \mathfrak{gl}(n, \mathbb{K})$, *consider the map* $\mathbb{K} \to \mathrm{GL}(n, \mathbb{K}) \colon t \mapsto \exp(tx)$. *Then* $\exp((t + s)x) = \exp(tx) \exp(sx)$. *In other words, this map is a morphism of Lie groups.*
(5) *The exponential map agrees with change of basis and transposition:*
$\exp(AxA^{-1}) = A \exp(x)A^{-1}$, $\exp(x^t) = (\exp(x))^t$.

A full proof of this theorem will not be given here; instead, we just give a sketch. The first two statements are just equalities of formal power series in one variable; thus, it suffices to check that they hold for $x \in \mathbb{R}$. Similarly, the third one is an identity of formal power series in two commuting variables, so it again follows from the well-known equality for $x, y \in \mathbb{R}$. The fourth follows from the third, and the fifth follows from $(AxA^{-1})^n = Ax^nA^{-1}$ and $(A^t)^n = (A^n)^t$.

Note that group morphisms $\mathbb{K} \to G$ are frequently called *one-parameter subgroups* in G. Thus, we can reformulate part (4) of the theorem by saying that $\exp(tx)$ is a one-parameter subgroup in $\mathrm{GL}(n, \mathbb{K})$.

How does it help us to study various matrix groups? The key idea is that the logarithmic map identifies some neighborhood of the identity in $GL(n, \mathbb{K})$ with some neighborhood of 0 in the vector space $\mathfrak{gl}(n, \mathbb{K})$. It turns out that it also does the same for all of the classical groups.

Theorem 2.30. *For each classical group $G \subset GL(n, \mathbb{K})$, there exists a vector space $\mathfrak{g} \subset \mathfrak{gl}(n, \mathbb{K})$ such that for some some neighborhood U of 1 in $GL(n, \mathbb{K})$ and some neighborhood u of 0 in $\mathfrak{gl}(n, \mathbb{K})$ the following maps are mutually inverse*

$$(U \cap G) \underset{\exp}{\overset{\log}{\rightleftarrows}} (u \cap \mathfrak{g}).$$

Before proving this theorem, note that it immediately implies the following important corollary.

Corollary 2.31. *Each classical group is a Lie group, with tangent space at identity $T_1 G = \mathfrak{g}$ and $\dim G = \dim \mathfrak{g}$. Groups $U(n)$, $SU(n)$, $Sp(n)$ are real Lie groups; groups $GL(n, \mathbb{K})$, $SL(n, \mathbb{K})$, $SO(n, \mathbb{K})$, $O(n, \mathbb{K})$, $Sp(n, \mathbb{K})$ are real Lie groups for $\mathbb{K} = \mathbb{R}$ and complex Lie groups for $\mathbb{K} = \mathbb{C}$.*

Let us prove this corollary first because it is very easy. Indeed, Theorem 2.30 shows that near 1, G is identified with an open set in a vector space. So it is immediately apparent that near 1, G is locally a submanifold in $GL(n, \mathbb{K})$. If $g \in G$ then $g \cdot U$ is a neighborhood of g in $GL(n, \mathbb{K})$, and $(g \cdot U) \cap G = g \cdot (U \cap G)$ is a neighborhood of g in G; thus, G is a submanifold in a neighborhood of g.

For the second part, consider the differential of the exponential map $\exp_* : T_0 \mathfrak{g} \to T_1 G$. Since $\mathfrak{g}$ is a vector space, $T_0 \mathfrak{g} = \mathfrak{g}$, and since $\exp(x) = 1 + x + \cdots$, the derivative is the identity; thus, $T_0 \mathfrak{g} = \mathfrak{g} = T_1 G$.

Proof of Theorem 2.30. The proof is case by case; it can not be any other way, as "classical groups" are defined by a list rather than by some general definition.

$GL(n, \mathbb{K})$: Immediate from Theorem 2.29.; in this case, $\mathfrak{g} = \mathfrak{gl}(n, \mathbb{K})$ is the space of all matrices.

$SL(n, \mathbb{K})$: Suppose $X \in SL(n, \mathbb{K})$ is close enough to identity. Then $X = \exp(x)$ for some $x \in \mathfrak{gl}(n, \mathbb{K})$. The condition that $X \in SL(n, \mathbb{K})$ is equivalent to $\det X = 1$, or $\det \exp(x) = 1$. But it is well-known that $\det \exp(x) = \exp(\operatorname{tr}(x))$ (which is easy to see by finding a basis in which x is upper-triangular), so $\exp(x) \in SL(n, \mathbb{K})$ if and only if $\operatorname{tr}(x) = 0$. Thus, in this case the statement also holds, with $\mathfrak{g} = \{x \in \mathfrak{gl}(n, \mathbb{K}) \mid \operatorname{tr} x = 0\}$.

O(n, $\mathbb{K}$), SO(n, $\mathbb{K}$): The group O(n, $\mathbb{K}$) is defined by $XX^t = I$. Then X, X^t commute. Writing $X = \exp(x), X^t = \exp(x^t)$ (since the exponential map agrees with transposition), we see that x, x^t also commute, and thus $\exp(x) \in$ O(n, $\mathbb{K}$) implies $\exp(x)\exp(x^t) = \exp(x + x^t) = 1$, so $x + x^t = 0$; conversely, if $x + x^t = 0$, then x, x^t commute, so we can reverse the argument to get $\exp(x) \in$ O(n, $\mathbb{K}$). Thus, in this case the theorem also holds, with $\mathfrak{g} = \{x \mid x + x^t = 0\}$ – the space of skew-symmetric matrices.

What about SO(n, $\mathbb{K}$)? In this case, we should add to the condition $XX^t = 1$ (which gives $x + x^t = 0$) also the condition $\det X = 1$, which gives $\mathrm{tr}(x) = 0$. However, this last condition is unnecessary, because $x + x^t = 0$ implies that all diagonal entries of x are zero. So both O(n, $\mathbb{K}$) and SO(n, $\mathbb{K}$) correspond to the same space of matrices $\mathfrak{g} = \{x \mid x + x^t = 0\}$. This might seem confusing until one realizes that SO(n, $\mathbb{K}$) is exactly the connected component of identity in O(n, $\mathbb{K}$); thus, a neighborhood of 1 in O(n, $\mathbb{K}$) coincides with a neighborhood of 1 in SO(n, $\mathbb{K}$).

U(n), SU(n): A similar argument shows that for x in a neighborhood of the identity in $\mathfrak{gl}(n, \mathbb{C})$, $\exp x \in$ U(n) $\iff x + x^* = 0$ (where $x^* = \bar{x}^t$) and $\exp x \in$ SU(n) $\iff x + x^* = 0, \mathrm{tr}(x) = 0$. Note that in this case, $x + x^*$ does not imply that x has zeroes on the diagonal: it only implies that the diagonal entries are purely imaginary. Thus, $\mathrm{tr}\, x = 0$ does not follow automatically from $x + x^* = 0$, so in this case the tangent spaces for U(n), SU(n) are different.

Sp(n, $\mathbb{K}$): A similar argument shows that $\exp(x) \in$ Sp(n, $\mathbb{K}$) $\iff x + J^{-1}x^t J = 0$ where J is given by (2.5). Thus, in this case the theorem also holds.

Sp(n): The same arguments as above show that $\exp(x) \in$ Sp(n) $\iff x + J^{-1}x^t J = 0, x + x^* = 0$. $\qquad\square$

The vector space $\mathfrak{g} = T_1 G$ is called the *Lie algebra* of the corresponding group G (this will be justified later, when we actually define an algebra operation on it). Traditionally, the Lie algebra is denoted by lowercase letters using Fraktur (Old German) fonts: for example, the Lie algebra of group SU(n) is denoted by $\mathfrak{su}(n)$.

Theorem 2.30 gives "local" information about classical Lie groups, i.e. the description of the tangent space at identity. In many cases, it is also important to know "global" information, such as the topology of the group G. In some low-dimensional cases, it is possible to describe the topology of G by establishing a diffeomorphism of G with a known manifold. For example, we have shown in Example 2.5 that SU(2) $\simeq S^3$; it is shown in Exercise 2.10 that SO(3, $\mathbb{R}$) $\simeq$

$SU(2)/\mathbb{Z}_2$ and thus is diffeomorphic to the real projective space $\mathbb{RP}^3$. For higher dimensional groups, the standard method of finding their topological invariants such as fundamental groups is by using the results of Corollary 2.12: if G acts transitively on a manifold M, then G is a fiber bundle over M with the fiber G_m–stabilizer of point in M. Thus we can get information about fundamental groups of G from fundamental groups of M, G_m. Details of this approach for different classical groups are given in the exercises (see Exercises 2.11, 2.12, and 2.16).

Tables 2.1, 2.2, and 2.3 summarize the results of Theorem 2.30 and the computation of the fundamental groups of classical Lie groups given in the exercises. For nonconnected groups, $\pi_1(G)$ stands for the fundamental group of the connected component of identity.

For complex classical groups, the Lie algebra and dimension are given by the same formula as for real groups. However, the topology of complex Lie groups is different and is given in Table 2.3. We do not give a proof of these results, referring the reader to more advanced books such as [32].

Table 2.1. *Compact classical groups. Here π_0 is the set of connected components, π_1 is the fundamental group (for disconnected groups, π_1 is the fundamental group of the connected component of identity), and J is given by (2.5).*

G	$O(n, \mathbb{R})$	$SO(n, \mathbb{R})$	$U(n)$	$SU(n)$	$Sp(n)$
$\mathfrak{g}$	$x + x^t = 0$	$x + x^t = 0$	$x + x^* = 0$	$x + x^* = 0$, $\operatorname{tr} x = 0$	$x + J^{-1}x^t J = 0$ $x + x^* = 0$
$\dim G$	$\frac{n(n-1)}{2}$	$\frac{n(n-1)}{2}$	n^2	$n^2 - 1$	$n(2n + 1)$
$\pi_0(G)$	$\mathbb{Z}_2$	$\{1\}$	$\{1\}$	$\{1\}$	$\{1\}$
$\pi_1(G)$	$\mathbb{Z}_2 \ (n \geq 3)$	$\mathbb{Z}_2 \ (n \geq 3)$	$\mathbb{Z}$	$\{1\}$	$\{1\}$

Table 2.2. *Noncompact real classical groups.*

G	$GL(n, \mathbb{R})$	$SL(n, \mathbb{R})$	$Sp(n, \mathbb{R})$
$\mathfrak{g}$	$\mathfrak{gl}(n, \mathbb{R})$	$\operatorname{tr} x = 0$	$x + J^{-1}x^t J = 0$
$\dim G$	n^2	$n^2 - 1$	$n(2n + 1)$
$\pi_0(G)$	$\mathbb{Z}_2$	$\{1\}$	$\{1\}$
$\pi_1(G)$	$\mathbb{Z}_2 \ (n \geq 3)$	$\mathbb{Z}_2 \ (n \geq 3)$	$\mathbb{Z}$

Table 2.3. *Complex classical groups.*

G	$GL(n, \mathbb{C})$	$SL(n, \mathbb{C})$	$O(n, \mathbb{C})$	$SO(n, \mathbb{C})$
$\pi_0(G)$	$\{1\}$	$\{1\}$	$\mathbb{Z}_2$	$\{1\}$
$\pi_1(G)$	$\mathbb{Z}$	$\{1\}$	$\mathbb{Z}_2$	$\mathbb{Z}_2$

Note that some of the classical groups are not simply-connected. As was shown in Theorem 2.7, in this case the universal cover has a canonical structure of a Lie group. Of special importance is the universal cover of $SO(n, \mathbb{R})$ which is called the *spin group* and is denoted $\mathrm{Spin}(n)$; since $\pi_1(SO(n, \mathbb{R})) = \mathbb{Z}_2$, this is a twofold cover, so $\mathrm{Spin}(n)$ is a compact Lie group.

2.8. Exercises

2.1. Let G be a Lie group and H – a closed Lie subgroup.
 (1) Let $\overline{H}$ be the closure of H in G. Show that $\overline{H}$ is a subgroup in G.
 (2) Show that each coset $Hx, x \in \overline{H}$, is open and dense in $\overline{H}$.
 (3) Show that $\overline{H} = H$, that is, every closed Lie subgroup is indeed a closed subset in G.

2.2. (1) Show that every discrete normal subgroup of a connected Lie group is central (hint: consider the map $G \to N : g \mapsto ghg^{-1}$ where h is a fixed element in N).
 (2) By applying part (a) to kernel of the map $\widetilde{G} \to G$, show that for any connected Lie group G, the fundamental group $\pi_1(G)$ is commutative.

2.3. Let $f : G_1 \to G_2$ be a morphism of connected Lie groups such that $f_* : T_1G_1 \to T_1G_2$ is an isomorphism (such a morphism is sometimes called *local isomorphism*). Show that f is a covering map, and $\mathrm{Ker} f$ is a discrete central subgroup.

2.4. Let $\mathcal{F}_n(\mathbb{C})$ be the set of all flags in $\mathbb{C}^n$ (see Example 2.25). Show that

$$\mathcal{F}_n(\mathbb{C}) = GL(n, \mathbb{C})/B(n, \mathbb{C}) = U(n)/T(n)$$

where $B(n, \mathbb{C})$ is the group of invertible complex upper triangular matrices, and $T(n)$ is the group of diagonal unitary matrices (which is easily shown to be the n-dimensional torus $(\mathbb{R}/\mathbb{Z})^n$). Deduce from this that $\mathcal{F}_n(\mathbb{C})$ is a compact complex manifold and find its dimension over $\mathbb{C}$.

2.5. Let $G_{n,k}$ be the set of all dimension k subspaces in $\mathbb{R}^n$ (usually called the Grassmanian). Show that $G_{n,k}$ is a homogeneous space for the group $O(n, \mathbb{R})$ and thus can be identified with coset space $O(n, \mathbb{R})/H$ for appropriate H. Use it to prove that $G_{n,k}$ is a manifold and find its dimension.

2.6. Show that if $G = GL(n, \mathbb{R}) \subset \text{End}(\mathbb{R}^n)$ so that each tangent space is canonically identified with $\text{End}(\mathbb{R}^n)$, then $(L_g)_* v = gv$ where the product in the right-hand side is the usual product of matrices, and similarly for the right action. Also, the adjoint action is given by $\text{Ad } g(v) = gvg^{-1}$.

Exercises 2.7–2.10 are about the group SU(2) and its adjoint representation

2.7. Define a bilinear form on $\mathfrak{su}(2)$ by $(a, b) = \frac{1}{2} \text{tr}(a\overline{b}^t)$. Show that this form is symmetric, positive definite, and invariant under the adjoint action of SU(2).

2.8. Define a basis in $\mathfrak{su}(2)$ by

$$i\sigma_1 = \begin{pmatrix} 0 & i \\ i & 0 \end{pmatrix} \qquad i\sigma_2 = \begin{pmatrix} 0 & 1 \\ -1 & 0 \end{pmatrix} \qquad i\sigma_3 = \begin{pmatrix} i & 0 \\ 0 & -i \end{pmatrix}$$

Show that the map

$$\varphi \colon SU(2) \to GL(3, \mathbb{R}) \tag{2.8}$$

$$g \mapsto \text{matrix of Ad } g \text{ in the basis } i\sigma_1, i\sigma_2, i\sigma_3$$

gives a morphism of Lie groups $SU(2) \to SO(3, \mathbb{R})$.

2.9. Let $\varphi \colon SU(2) \to SO(3, \mathbb{R})$ be the morphism defined in the previous problem. Compute explicitly the map of tangent spaces $\varphi_* \colon \mathfrak{su}(2) \to \mathfrak{so}(3, \mathbb{R})$ and show that φ_* is an isomorphism. Deduce from this that Ker φ is a discrete normal subgroup in SU(2), and that Im φ is an open subgroup in $SO(3, \mathbb{R})$.

2.10. Prove that the map φ used in two previous exercises establishes an isomorphism $SU(2)/\mathbb{Z}_2 \to SO(3, \mathbb{R})$ and thus, since $SU(2) \simeq S^3$, $SO(3, \mathbb{R}) \simeq \mathbb{RP}^3$.

2.11. Using Example 2.24, show that for $n \geq 1$, we have $\pi_0(SU(n + 1)) = \pi_0(SU(n))$, $\pi_0(U(n + 1)) = \pi_0(U(n))$ and deduce from it that groups $U(n)$, $SU(n)$ are connected for all n. Similarly, show that for $n \geq 2$, we have $\pi_1(SU(n+1)) = \pi_1(SU(n))$, $\pi_1(U(n+1)) = \pi_1(U(n))$ and deduce from it that for $n \geq 2$, $SU(n)$ is simply-connected and $\pi_1(U(n)) = \mathbb{Z}$.

2.12. Using Example 2.24, show that for $n \geq 2$, we have $\pi_0(SO(n+1, \mathbb{R})) = \pi_0(SO(n, \mathbb{R}))$ and deduce from it that groups $SO(n)$ are connected for all $n \geq 2$. Similarly, show that for $n \geq 3$, $\pi_1(SO(n+1, \mathbb{R})) = \pi_1(SO(n, \mathbb{R}))$ and deduce from it that for $n \geq 3$, $\pi_1(SO(n, \mathbb{R})) = \mathbb{Z}_2$.

2.13. Using the Gram–Schmidt orthogonalization process, show that $GL(n, \mathbb{R})/O(n, \mathbb{R})$ is diffeomorphic to the space of upper-triangular matrices with positive entries on the diagonal. Deduce from this that $GL(n, \mathbb{R})$ is homotopic (as a topological space) to $O(n, \mathbb{R})$.

2.14. Let L_n be the set of all Lagrangian subspaces in $\mathbb{R}^{2n}$ with the standard symplectic form ω defined in Section 2.7. (A subspace V is Lagrangian if $\dim V = n$ and $\omega(x, y) = 0$ for any $x, y \in V$.)

Show that the group $Sp(n, \mathbb{R})$ acts transitively on L_n and use it to define on L_n a structure of a smooth manifold and find its dimension.

2.15. Let $\mathbb{H} = \{a + bi + cj + dk \mid a, b, c, d \in \mathbb{R}\}$ be the algebra of quaternions, defined by $ij = k = -ji, jk = i = -kj, ki = j = -ik, i^2 = j^2 = k^2 = -1$, and let $\mathbb{H}^n = \{(h_1, \ldots, h_n) \mid h_i \in \mathbb{H}\}$. In particular, the subalgebra generated by $1, i$ coincides with the field $\mathbb{C}$ of complex numbers.

Note that $\mathbb{H}^n$ has a structure of both left and right module over $\mathbb{H}$ defined by

$$h(h_1, \ldots, h_n) = (hh_1, \ldots, hh_n), \qquad (h_1, \ldots, h_n)h = (h_1 h, \ldots, h_n h)$$

(1) Let $\mathrm{End}_{\mathbb{H}}(\mathbb{H}^n)$ be the algebra of endomorphisms of $\mathbb{H}^n$ considered as right $\mathbb{H}$-module:

$$\mathrm{End}_{\mathbb{H}}(\mathbb{H}^n) = \{A \colon \mathbb{H}^n \to \mathbb{H}^n \mid A(\mathbf{h} + \mathbf{h}')$$
$$= A(\mathbf{h}) + A(\mathbf{h}'),\ A(\mathbf{h}h) = A(\mathbf{h})h\}$$

Show that $\mathrm{End}_{\mathbb{H}}(\mathbb{H}^n)$ is naturally identified with the algebra of $n \times n$ matrices with quaternion entries.

(2) Define an $\mathbb{H}$-valued form $(\ ,\ )$ on $\mathbb{H}^n$ by

$$(\mathbf{h}, \mathbf{h}') = \sum_i \overline{h_i} h_i'$$

where $\overline{a + bi + cj + dk} = a - bi - cj - dk$. (Note that $\overline{uv} = \overline{v}\,\overline{u}$.)
Let $U(n, \mathbb{H})$ be the group of "unitary quaternionic transformations":

$$U(n, \mathbb{H}) = \{A \in \mathrm{End}_{\mathbb{H}}(\mathbb{H}^n) \mid (A\mathbf{h}, A\mathbf{h}') = (\mathbf{h}, \mathbf{h}')\}.$$

Show that this is indeed a group and that a matrix A is in $U(n, \mathbb{H})$ iff $A^*A = 1$, where $(A^*)_{ij} = \overline{A_{ji}}$.

(3) Define a map $\mathbb{C}^{2n} \simeq \mathbb{H}^n$ by

$$(z_1, \ldots, z_{2n}) \mapsto (z_1 + jz_{n+1}, \ldots, z_n + jz_{2n})$$

Show that it is an isomorphism of complex vector spaces (if we consider $\mathbb{H}^n$ as a complex vector space by $z(h_1, \ldots h_n) = (h_1 z, \ldots, h_n z)$) and that this isomorphism identifies

$$\mathrm{End}_{\mathbb{H}}(\mathbb{H}^n) = \{A \in \mathrm{End}_{\mathbb{C}}(\mathbb{C}^{2n}) \mid \overline{A} = J^{-1}AJ\}$$

where J is defined by (2.5). (Hint: use $jz = \overline{z}j$ for any $z \in \mathbb{C}$ to show that $\mathbf{h} \mapsto \mathbf{h}j$ is identified with $\mathbf{z} \mapsto J\overline{\mathbf{z}}$.)

(4) Show that under identification $\mathbb{C}^{2n} \simeq \mathbb{H}^n$ defined above, the quaternionic form $(\,,\,)$ is identified with

$$(\mathbf{z}, \mathbf{z}') - j\langle \mathbf{z}, \mathbf{z}' \rangle$$

where $(\mathbf{z}, \mathbf{z}') = \sum \overline{z_i} z_i'$ is the standard Hermitian form in $\mathbb{C}^{2n}$ and $\langle \mathbf{z}, \mathbf{z}' \rangle = \sum_{i=1}^{n}(z_{i+n} z_i' - z_i z_{i+n}')$ is the standard bilinear skew-symmetric form in $\mathbb{C}^{2n}$. Deduce from this that the group $U(n, \mathbb{H})$ is identified with $\mathrm{Sp}(n) = \mathrm{Sp}(n, \mathbb{C}) \cap SU(2n)$.

2.16. (1) Show that $\mathrm{Sp}(1) \simeq SU(2) \simeq S^3$.

(2) Using the previous exercise, show that we have a natural transitive action of $\mathrm{Sp}(n)$ on the sphere S^{4n-1} and a stabilizer of a point is isomorphic to $\mathrm{Sp}(n-1)$.

(3) Deduce that $\pi_1(\mathrm{Sp}(n+1)) = \pi_1(\mathrm{Sp}(n))$, $\pi_0(\mathrm{Sp}(n+1)) = \pi_0(\mathrm{Sp}(n))$.

3

Lie groups and Lie algebras

3.1. Exponential map

We are now turning to the study of arbitrary Lie groups. Our first goal will be to generalize the exponential map $\exp\colon \mathfrak{g} \to G$, $\mathfrak{g} = T_1 G$, which proved so useful in the study of matrix groups (see Theorem 2.29), to general Lie groups. We can not use power series to define it because we do not have multiplication in $\mathfrak{g}$. However, it turns out that there is still a way to define such a map so that most of the results about the exponential map for matrix groups can be generalized to arbitrary groups, and this gives us a key to studying Lie groups. This definition is based on the notion of a one-parameter subgroup (compare with Theorem 2.29).

Proposition 3.1. *Let G be a real or complex Lie group, $\mathfrak{g} = T_1 G$, and let $x \in \mathfrak{g}$. Then there exists a unique morphism of Lie groups $\gamma_x\colon \mathbb{K} \to G$ such that*

$$\dot{\gamma}_x(0) = x,$$

where dot stands for derivative with respect to t. The map γ_x will be called the one-parameter subgroup corresponding to x.

Proof. Let us first consider the case of a real Lie group. We begin with uniqueness. The usual argument, used to compute the derivative of e^x in calculus, shows that if $\gamma(t)$ is a one-parameter subgroup, then $\dot{\gamma}(t) = \gamma(t) \cdot \dot{\gamma}(0) = \dot{\gamma}(0) \cdot \gamma(t)$. This is immediate for matrix groups; for general groups, the same proof works if, as in Section 2.6, we interpret $\gamma(t) \cdot \dot{\gamma}(0)$ as $(L_{\gamma(t)})_* \dot{\gamma}(0)$ and similarly for the right action. This gives us a differential equation for γ: if v_x is a left-invariant vector field on G such that $v_x(1) = x$, then γ is an integral curve for v. This proves uniqueness of $\gamma_x(t)$.

For existence, let $\Phi^t\colon G \to G$ be the time t flow of the vector field v_x (*a priori*, it is only defined for small enough t). Since the vector field is left-invariant, the flow operator is also left-invariant: $\Phi^t(g_1g_2) = g_1\Phi^t(g_2)$. Now let $\gamma(t) = \Phi^t(1)$. Then $\gamma(t+s) = \Phi^{t+s}(1) = \Phi^s(\Phi^t(1)) = \Phi^s(\gamma(t) \cdot 1) = \gamma(t)\Phi^s(1) = \gamma(t)\gamma(s)$ as desired. This proves the existence of γ for small enough t. The fact that it can be extended to any $t \in \mathbb{R}$ is obvious from $\gamma(t+s) = \gamma(t)\gamma(s)$.

The proof for complex Lie groups is similar but uses generalization of the usual results of the theory of differential equations to complex setup (such as defining "time t flow" for complex time t). □

Note that a one-parameter subgroup may not be a closed Lie subgroup (as is easy to see from Example 2.13); however, it will always be a Lie subgroup in G.

Definition 3.2. Let G be a real or complex Lie group, $\mathfrak{g} = T_1G$. Then the exponential map $\exp\colon \mathfrak{g} \to G$ is defined by

$$\exp(x) = \gamma_x(1),$$

where $\gamma_x(t)$ is the one-parameter subgroup with tangent vector at 1 equal to x.

Note that the uniqueness of one-parameter subgroups immediately implies that $\gamma_x(\lambda t) = \gamma_{\lambda x}(t)$ for any $\lambda \in \mathbb{K}$. Indeed, $\gamma_x(\lambda t)$ is a one-parameter subgroup with $d\gamma_x(\lambda t)/dt|_{t=0} = \lambda x$. Thus, $\gamma_x(t)$ only depends on the product $tx \in \mathfrak{g}$, so

$$\gamma_x(t) = \gamma_{tx}(1) = \exp(tx).$$

Example 3.3. For $G \subset GL(n, \mathbb{K})$, it follows from Theorem 2.29 that this definition agrees with the exponential map defined by (2.6).

Example 3.4. Let $G = \mathbb{R}$, so that $\mathfrak{g} = \mathbb{R}$. Then for any $a \in \mathfrak{g}$, the corresponding one-parameter subgroup is $\gamma_a(t) = ta$, so the exponential map is given by $\exp(a) = a$.

Example 3.5. Let $G = S^1 = \mathbb{R}/\mathbb{Z} = \{z \in \mathbb{C} \mid |z| = 1\}$ (these two descriptions are related by $z = e^{2\pi i\theta}, \theta \in \mathbb{R}/\mathbb{Z}$). Then $\mathfrak{g} = \mathbb{R}$, and the exponential map is given by $\exp(a) = a \mod \mathbb{Z}$ (if we use $G = \mathbb{R}/\mathbb{Z}$ description) or $\exp(a) = e^{2\pi i a}$ (if we use $G = \{z \in \mathbb{C} \mid |z| = 1\}$).

Note that the construction of the one-parameter subgroup given in the proof of Proposition 3.1 immediately gives the following result, formal proof of which is left as an exercise to the reader.

Proposition 3.6.

(1) *Let v be a left-invariant vector field on G. Then the time t flow of this vector field is given by $g \mapsto g \exp(tx)$, where $x = v(1)$.*

(2) *Let v be a right-invariant vector field on G. Then the time t flow of this vector field is given by $g \mapsto \exp(tx)g$, where $x = v(1)$.*

The following theorem summarizes properties of the exponential map.

Theorem 3.7. *Let G be a real or complex Lie group and $\mathfrak{g} = T_1 G$.*

(1) $\exp(x) = 1 + x + \ldots$ *(that is, $\exp(0) = 1$ and $\exp_*(0) \colon \mathfrak{g} \to T_1 G = \mathfrak{g}$ is the identity map).*

(2) *The exponential map is a diffeomorphism (for complex G, invertible analytic map) between some neighborhood of 0 in $\mathfrak{g}$ and a neighborhood of 1 in G. The local inverse map will be denoted by* log.

(3) $\exp((t + s)x) = \exp(tx)\exp(sx)$ *for any $s, t \in \mathbb{K}$.*

(4) *For any morphism of Lie groups $\varphi \colon G_1 \to G_2$ and any $x \in \mathfrak{g}_1$, we have* $\exp(\varphi_*(x)) = \varphi(\exp(x))$.

(5) *For any $X \in G, y \in \mathfrak{g}$, we have $X \exp(y) X^{-1} = \exp(\mathrm{Ad}\, X . y)$, where* Ad *is the adjoint action of G on $\mathfrak{g}$ defined by* (2.4).

Proof. The first statement is immediate from the definition. Differentiability (respectively, analyticity) of exp follows from the construction of γ_x given in the proof of Proposition 3.1 and general results about the dependence of a solution of a differential equation on initial condition. The fact that exp is locally invertible follows from (1) and inverse function theorem.

The third statement is again an immediate corollary of the definition ($\exp(tx)$ is a one-parameter subgroup in G).

Statement 4 follows from the uniqueness of one-parameter subgroup. Indeed, $\varphi(\exp(tx))$ is a one-parameter subgroup in G_2 with tangent vector at identity $\varphi_*(\exp_*(x)) = \varphi_*(x)$. Thus, $\varphi(\exp(tx)) = \exp(t\varphi_*(x))$.

The last statement is a special case of the previous one: the map $Y \mapsto XYX^{-1}$ is a morphism of Lie groups $G \to G$. □

Comparing this with Theorem 2.29, we see that we have many of the same results. A notable exception is that we have no analog of the statement that if $xy = yx$, then $\exp(x)\exp(y) = \exp(y)\exp(x)$. In fact the statement does not make sense for general groups, as the product xy is not defined. A proper analog of this statement will be proved later (Theorem 3.36).

Remark 3.8. In general, the exponential map is not surjective – see Exercise 3.1. However, it can be shown that for compact Lie groups, the exponential map is surjective.

Proposition 3.9. *Let G_1, G_2 be Lie groups (real or complex). If G_1 is connected, then any Lie group morphism $\varphi: G_1 \to G_2$ is uniquely determined by the linear map $\varphi_*: T_1G_1 \to T_1G_2$.*

Proof. By Theorem 3.7, $\varphi(\exp x) = \exp(\varphi_*(x))$. Since the image of the exponential map contains a neighborhood of identity in G_1, this implies that φ_* determines φ in a neighborhood of identity in G_1. But by Corollary 2.10, any neighborhood of the identity generates G_1.   □

Example 3.10. Let $G = SO(3, \mathbb{R})$. Then $T_1G = \mathfrak{so}(3, \mathbb{R})$ consists of skew-symmetric 3×3 matrices. One possible choice of a basis in $\mathfrak{so}(3, \mathbb{R})$ is

$$J_x = \begin{pmatrix} 0 & 0 & 0 \\ 0 & 0 & -1 \\ 0 & 1 & 0 \end{pmatrix}, \qquad J_y = \begin{pmatrix} 0 & 0 & 1 \\ 0 & 0 & 0 \\ -1 & & 0 \end{pmatrix}, \qquad J_z = \begin{pmatrix} 0 & -1 & 0 \\ 1 & 0 & 0 \\ 0 & 0 & 0 \end{pmatrix}$$
(3.1)

We can explicitly describe the corresponding subgroups in G. Namely,

$$\exp(tJ_x) = \begin{pmatrix} 1 & 0 & 0 \\ 0 & \cos t & -\sin t \\ 0 & \sin t & \cos t \end{pmatrix}$$

is rotation around x-axis by angle t; similarly, J_y, J_z generate rotations around y, z axes. The easiest way to show this is to note that such rotations do form a one-parameter subgroup; thus, they must be of the form $\exp(tJ)$ for some $J \in \mathfrak{so}(3, \mathbb{R})$, and then compute the derivative to find J.

By Theorem 3.7, elements of the form $\exp(tJ_x)$, $\exp(tJ_y)$, $\exp(tJ_z)$ generate a neighborhood of identity in $SO(3, \mathbb{R})$. Since $SO(3, \mathbb{R})$ is connected, by Corollary 2.10, these elements generate the whole group $SO(3, \mathbb{R})$. For this reason, it is common to refer to J_x, J_y, J_z as "infinitesimal generators" of $SO(3, \mathbb{R})$. Thus, in a certain sense $SO(3, \mathbb{R})$ is generated by three elements.

3.2. The commutator

So far, we have considered $\mathfrak{g} = T_1G$ as a vector space with no additional structure. However, since the exponential map locally identifies G with $\mathfrak{g}$, the multiplication in G defines a certain operation in $\mathfrak{g}$. Namely, for sufficiently small $x, y \in \mathfrak{g}$, the product $\exp(x)\exp(y)$ will be close to $1 \in G$ and thus can be written in the form

$$\exp(x)\exp(y) = \exp(\mu(x, y))$$

for some smooth (for complex Lie groups, complex analytic) map $\mu\colon \mathfrak{g} \times \mathfrak{g} \to \mathfrak{g}$ defined in a neighborhood of $(0, 0)$. The map μ is sometimes called the *group law in logarithmic coordinates*.

Lemma 3.11. *The Taylor series for μ is given by*

$$\mu(x, y) = x + y + \lambda(x, y) + \cdots$$

where dots stand for the terms of order ≥ 3 and $\lambda\colon \mathfrak{g} \times \mathfrak{g} \to \mathfrak{g}$ is a bilinear skew-symmetric (that is, satisfying $\lambda(x, y) = -\lambda(y, x)$) map.

Proof. Any smooth map can be written in the form $\alpha_1(x) + \alpha_2(y) + Q_1(x) + Q_2(y) + \lambda(x, y) + \cdots$, where α_1, α_2 are linear maps $\mathfrak{g} \to \mathfrak{g}$, Q_1, Q_2 are quadratic, and λ is bilinear. Letting $y = 0$, we see that $\mu(x, 0) = x$, which gives $\alpha_1(x) = x, Q_1(x) = 0$; similar argument shows that $\alpha_2(y) = y, Q_2(y) = 0$. Thus, $\mu(x, y) = x + y + \lambda(x, y) + \cdots$.

To show that λ is skew-symmetric, it suffices to check that $\lambda(x, x) = 0$. But $\exp(x)\exp(x) = \exp(2x)$, so $\mu(x, x) = x + x$. □

For reasons that will be clear in the future, it is traditional to introduce notation $[x, y] = 2\lambda(x, y)$, so we have

$$\exp(x)\exp(y) = \exp(x + y + \frac{1}{2}[x, y] + \cdots) \tag{3.2}$$

for some bilinear skew-symmetric map $[\,,\,]\colon \mathfrak{g} \times \mathfrak{g} \to \mathfrak{g}$. This map is called the *commutator*.

Thus, we see that for any Lie group, its tangent space at identity $\mathfrak{g} = T_1 G$ has a canonical skew-symmetric bilinear operation, which appears as the lowest non-trivial term of the Taylor series for multiplication in G. This operation has the following properties.

Proposition 3.12.

(1) *Let $\varphi\colon G_1 \to G_2$ be a morphism of real or complex Lie groups and $\varphi_*\colon \mathfrak{g}_1 \to \mathfrak{g}_2$, where $\mathfrak{g}_1 = T_1 G_1$, $\mathfrak{g}_2 = T_1 G_2$ – the corresponding map of tangent spaces at identity. Then φ_* preserves the commutator:*

$$\varphi_*[x, y] = [\varphi_* x, \varphi_* y] \qquad \text{for any } x, y \in \mathfrak{g}_1.$$

(2) *The adjoint action of a Lie group G on $\mathfrak{g} = T_1 G$ preserves the commutator:*
    $\operatorname{Ad} g([x, y]) = [\operatorname{Ad} g.x, \operatorname{Ad} g.y]$ *for any $x, y \in \mathfrak{g}$.*

(3)

$$\exp(x)\exp(y)\exp(-x)\exp(-y) = \exp([x, y] + \cdots), \tag{3.3}$$

where dots stand for terms of degree three and higher.

Proof. The first statement is immediate from the definition of commutator (3.2) and the fact that every morphism of Lie groups commutes with the exponential map (Theorem 3.7). The second follows from the first and the fact that for any $g \in G$, the map $\mathrm{Ad}\, g \colon G \to G$ is a morphism of Lie groups.

The last formula is proved by explicit computation using (3.2).   □

This theorem shows that the commutator in $\mathfrak{g}$ is closely related with the group commutator in G, which explains the name.

Corollary 3.13. *If G is a commutative Lie group, then $[x, y] = 0$ for all $x, y \in \mathfrak{g}$.*

Example 3.14. Let $G \subset \mathrm{GL}(n, \mathbb{K})$, so that $\mathfrak{g} \subset \mathfrak{gl}(n, \mathbb{K})$. Then the commutator is given by $[x, y] = xy - yx$. Indeed, using (3.3) and keeping only linear and bilinear terms, we can write $(1+x+\cdots)(1+y+\cdots)(1-x+\cdots)(1-y+\cdots) = 1 + [x, y] + \cdots$ which gives $[x, y] = xy - yx$.

3.3. Jacobi identity and the definition of a Lie algebra

So far, for a Lie group G, we have defined a bilinear operation on $\mathfrak{g} = T_1 G$, which is obtained from the multiplication on G. An obvious question is whether the associativity of multiplication gives some identities for the commutator. In this section we will answer this question; as one might expect, the answer is "yes".

By results of Proposition 3.12, any morphism φ of Lie groups gives rise to a map φ_* of corresponding tangent spaces at identity which preserves the commutator. Let us apply it to the adjoint action defined in Section 2.6, which can be considered as a morphism of Lie groups

$$\mathrm{Ad} \colon G \to \mathrm{GL}(\mathfrak{g}). \tag{3.4}$$

Lemma 3.15. *Denote by* $\mathrm{ad} = \mathrm{Ad}_* \colon \mathfrak{g} \to \mathfrak{gl}(\mathfrak{g})$ *the map of tangent spaces corresponding to the map* (3.4). *Then*

(1) $\mathrm{ad}\, x.y = [x, y]$
(2) $\mathrm{Ad}(\exp x) = \exp(\mathrm{ad}\, x)$ *as operators* $\mathfrak{g} \to \mathfrak{g}$.

Proof. By definition of Ad, we have $\mathrm{Ad}\, g.y = \dfrac{d}{dt}|_{t=0}\big(g \exp(ty)g^{-1}\big)$. Thus, we see that ad is defined by

$$\mathrm{ad}\, x.y = \frac{d}{ds}\frac{d}{dt}\exp(sx)\exp(ty)\exp(-sx)|_{t=s=0}.$$

On the other hand, by (3.3), $\exp(sx)\exp(ty)\exp(-sx) = \exp(ty + ts[x, y] + \cdots)$. Combining these two results, we see that $\operatorname{ad} x.y = [x, y]$.

The second part is immediate from Theorem 3.7. □

Theorem 3.16. *Let G be a real or complex Lie group, $\mathfrak{g} = T_1G$ and let the commutator $[\,,\,]\colon \mathfrak{g} \times \mathfrak{g} \to \mathfrak{g}$ be defined by (3.2). Then it satisfies the following identity, called* Jacobi identity:

$$[x, [y, z]] = [[x, y], z] + [y, [x, z]]. \tag{3.5}$$

This identity can also be written in any of the following equivalent forms:

$$[x, [y, z]] + [y, [z, x]] + [z, [x, y]] = 0$$
$$\operatorname{ad} x.[y, z] = [\operatorname{ad} x.y, z] + [y, \operatorname{ad} x.z] \tag{3.6}$$
$$\operatorname{ad}[x, y] = \operatorname{ad} x \operatorname{ad} y - \operatorname{ad} y \operatorname{ad} x.$$

Proof. Since Ad is a morphism of Lie groups $G \to GL(\mathfrak{g})$, by Proposition 3.12, $\operatorname{ad}\colon \mathfrak{g} \to \mathfrak{gl}(\mathfrak{g})$ must preserve commutator. But the commutator in $\mathfrak{gl}(\mathfrak{g})$ is given by $[A, B] = AB - BA$ (see Example 3.14), so $\operatorname{ad}[x, y] = \operatorname{ad} x \operatorname{ad} y - \operatorname{ad} y \operatorname{ad} x$, which proves the last formula of (3.6).

Equivalence of all forms of Jacobi identity is left as an exercise to the reader (see Exercise 3.3). □

Definition 3.17. A Lie algebra over a field $\mathbb{K}$ is a vector space $\mathfrak{g}$ over $\mathbb{K}$ with a $\mathbb{K}$-bilinear map $[\,,\,]\colon \mathfrak{g} \times \mathfrak{g} \to \mathfrak{g}$ which is skew-symmetric: $[x, y] = -[y, x]$ and satisfies Jacobi identity (3.5).

A morphism of Lie algebras is a $\mathbb{K}$-linear map $f\colon \mathfrak{g}_1 \to \mathfrak{g}_2$ which preserves the commutator.

This definition makes sense for any field; however, in this book we will only consider real ($\mathbb{K} = \mathbb{R}$) and complex ($\mathbb{K} = \mathbb{C}$) Lie algebras.

Example 3.18. Let $\mathfrak{g}$ be a vector space with the commutator defined by $[x, y] = 0$ for all $x, y \in \mathfrak{g}$. Then $\mathfrak{g}$ is a Lie algebra; such a Lie algebra is called *commutative*, or *abelian*, Lie algebra. This is motivated by Corollary 3.13, where it was shown that for a commutative Lie group G, $\mathfrak{g} = T_1G$ is naturally a commutative Lie algebra.

Example 3.19. Let A be an associative algebra over $\mathbb{K}$. Then the formula

$$[x, y] = xy - yx$$

defines on A a structure of a Lie algebra, which can be checked by a direct computation.

Using the notion of a Lie algebra, we can summarize much of the results of the previous two sections in the following theorem.

Theorem 3.20. *Let G be a real or complex Lie group. Then $\mathfrak{g} = T_1 G$ has a canonical structure of a Lie algebra over $\mathbb{K}$ with the commutator defined by (3.2); we will denote this Lie algebra by* Lie(G).

Every morphism of Lie groups $\varphi \colon G_1 \to G_2$ defines a morphism of Lie algebras $\varphi_* \colon \mathfrak{g}_1 \to \mathfrak{g}_2$, so we have a map Hom$(G_1, G_2) \to$ Hom$(\mathfrak{g}_1, \mathfrak{g}_2)$; *if G_1 is connected, then this map is injective:* Hom$(G_1, G_2) \subset$ Hom$(\mathfrak{g}_1, \mathfrak{g}_2)$.

3.4. Subalgebras and ideals

In the previous section, we have shown that for every Lie group G the vector space $\mathfrak{g} = T_1 G$ has a canonical structure of a Lie algebra, and every morphism of Lie groups gives rise to a morphism of Lie algebras.

Continuing the study of this correspondence between groups and algebras, we define analogs of Lie subgroups and normal subgroups.

Definition 3.21. Let $\mathfrak{g}$ be a Lie algebra over $\mathbb{K}$. A subspace $\mathfrak{h} \subset \mathfrak{g}$ is called a *Lie subalgebra* if it is closed under commutator, i.e. for any $x, y \in \mathfrak{h}$, we have $[x, y] \in \mathfrak{h}$. A subspace $\mathfrak{h} \subset \mathfrak{g}$ is called an *ideal* if for any $x \in \mathfrak{g}, y \in \mathfrak{h}$, we have $[x, y] \in \mathfrak{h}$.

It is easy to see that if $\mathfrak{h}$ is an ideal, then $\mathfrak{g}/\mathfrak{h}$ has a canonical structure of a Lie algebra.

Theorem 3.22. *Let G be a real or complex Lie group with Lie algebra $\mathfrak{g}$.*

(1) *Let H be a Lie subgroup in G (not necessarily closed). Then $\mathfrak{h} = T_1 H$ is a Lie subalgebra in $\mathfrak{g}$.*

(2) *Let H be a normal closed Lie subgroup in G. Then $\mathfrak{h} = T_1 H$ is an ideal in $\mathfrak{g}$, and* Lie$(G/H) = \mathfrak{g}/\mathfrak{h}$.

   Conversely, if H is a closed Lie subgroup in G, such that H, G are connected and $\mathfrak{h} = T_1 H$ is an ideal in $\mathfrak{g}$, then H is normal.

Proof. It easily follows from uniqueness statement for one-parameter subgroups that if $x \in T_1 H$, then $\exp(tx) \in H$ for all $t \in \mathbb{K}$. Using formula (3.3) for commutator, we see that for $x, y \in \mathfrak{h}$, left-hand side is in H; thus, $[x, y]$ must be in $T_1 H = \mathfrak{h}$.

Similarly, if H is a normal subgroup, then $\exp(x)\exp(y)\exp(-x) \in H$ for any $x \in \mathfrak{g}, y \in \mathfrak{h}$, so the left-hand side of (3.3) is again in H. Identity $\mathrm{Lie}(G/H) = \mathfrak{g}/\mathfrak{h}$ follows from Theorem 2.11.

Finally, if $\mathfrak{h}$ is an ideal in $\mathfrak{g}$, then it follows from $\mathrm{Ad}(\exp(x)) = \exp(\mathrm{ad}\,x)$ (Lemma 3.15) that for any $x \in \mathfrak{g}$, $\mathrm{Ad}(\exp(x))$ preserves $\mathfrak{h}$. Since expressions of the form $\exp(x), x \in \mathfrak{g}$, generate G (Corollary 2.10), this shows that for any $g \in G$, $\mathrm{Ad}\,g$ preserves $\mathfrak{h}$. Since by Theorem 3.7,

$$g\exp(y)g^{-1} = \exp(\mathrm{Ad}\,g.y), \quad g \in G, y \in \mathfrak{g},$$

we see that for any $y \in \mathfrak{h}$, $g\exp(y)g^{-1} \in H$. Since expressions $\exp y, y \in \mathfrak{h}$, generate H, we see that $ghg^{-1} \in H$ for any $h \in H$. $\square$

3.5. Lie algebra of vector fields

In this section, we illustrate the theory developed above in the example of the group $\mathrm{Diff}(M)$ of diffeomorphisms of a manifold M. For simplicity, throughout this section we only consider the case of real manifolds; however, all results also hold for complex manifolds.

The group $\mathrm{Diff}(M)$ is not a Lie group (it is infinite-dimensional), but in many ways it is similar to Lie groups. For example, it is easy to define what a smooth map from some group G to $\mathrm{Diff}(M)$ is: it is the same as an action of G on M by diffeomorphisms. Ignoring the technical problem with infinite-dimensionality for now, let us try to see what is the natural analog of the Lie algebra for the group $\mathrm{Diff}(M)$. It should be the tangent space at the identity; thus, its elements are derivatives of one-parameter families of diffeomorphisms.

Let $\varphi^t \colon M \to M$ be a one-parameter family of diffeomorphisms. Then, for every point $m \in M$, $\varphi^t(m)$ is a curve in M and thus $\frac{d}{dt}\varphi^t(m) \in T_m M$ is a tangent vector to M at m. In other words, $\frac{d}{dt}\varphi^t$ is a vector field on M. Thus, it is natural to define the Lie algebra of $\mathrm{Diff}(M)$ to be the space $\mathrm{Vect}(M)$ of all smooth vector fields on M.

What is the exponential map? If $\xi \in \mathrm{Vect}(M)$ is a vector field, then $\exp(t\xi)$ should be a one-parameter family of diffeomorphisms whose derivative is vector field ξ. So this is the solution of the differential equation

$$\frac{d}{dt}\varphi^t(m)|_{t=0} = \xi(m).$$

In other words, φ^t is the time t flow of the vector field ξ. We will denote it by

$$\exp(t\xi) = \Phi_\xi^t. \tag{3.7}$$

This may not be defined globally, but for the moment, let us ignore this problem.

What is the commutator $[\xi, \eta]$? By (3.3), we need to consider $\Phi_\xi^t \Phi_\eta^s \Phi_{-\xi}^t \Phi_{-\eta}^s$. It is well-known that this might not be the identity (if a plane flies 500 miles north, then 500 miles west, then 500 miles south, then 500 miles east, then it does not necessarily lands at the same spot it started – because Earth is not flat). By analogy with (3.3), we expect that this expression can be written in the form $1 + ts[\xi, \eta] + \cdots$ for some vector field $[\xi, \eta]$. This is indeed so, as the following proposition shows.

Proposition 3.23.

(1) *Let $\xi, \eta \in \mathrm{Vect}(M)$ be vector fields on M. Then there exists a unique vector field which we will denote by $[\xi, \eta]$ such that*

$$\Phi_\xi^t \Phi_\eta^s \Phi_{-\xi}^t \Phi_{-\eta}^s = \Phi_{[\xi,\eta]}^{ts} + \cdots, \qquad (3.8)$$

where dots stand for the terms of order 3 and higher in s, t.
(2) *The commutator (3.8) defines on the space of vector fields a structure of an (infinite-dimensional) real Lie algebra.*
(3) *The commutator can also be defined by any of the following formulas:*

$$[\xi, \eta] = \frac{\mathrm{d}}{\mathrm{d}t}(\Phi_\xi^t)_* \eta \qquad (3.9)$$

$$\partial_{[\xi,\eta]} f = \partial_\eta(\partial_\xi f) - \partial_\xi(\partial_\eta f), \qquad f \in C^\infty(M) \qquad (3.10)$$

$$\left[\sum f_i \partial_i, \sum g_j \partial_j\right] = \sum_{i,j} (g_i \partial_i(f_j) - f_i \partial_i(g_j)) \partial_j \qquad (3.11)$$

where $\partial_\xi(f)$ is the derivative of a function f in the direction of the vector field ξ, and $\partial_i = \frac{\partial}{\partial x^i}$ for some local coordinate system $\{x^i\}$.

The first two parts are, of course, to be expected, by analogy with finite-dimensional situation. However, since $\mathrm{Diff}(M)$ is not a finite-dimensional Lie group, we can not just refer to Theorem 3.20 but need to give a separate proof. Such a proof, together with the proof of the last part, can be found in any good book on differential geometry, for example in [49].

Remark 3.24. In many books the definition of commutator of vector fields differs by sign from the one given here. Both versions define on the space of vector fields a structure of Lie algebra, so it is a matter of choice which of the definitions to use. However, in our opinion the definition here – which naturally arises from the multiplication in the diffeomorphism group – is more natural, so we use it. Thus, when using results from other books, be sure to double-check which definition of commutator they use for vector fields.

The reason for the appearance of the minus sign is that the action of a diffeomorphism $\Phi: M \to M$ on functions on M is given by $(\Phi f)(m) = f(\Phi^{-1}m)$ (note the inverse!); thus, the derivative $\partial_\xi f = -\frac{d}{dt}\Phi_\xi^t f$. For example, if $\xi = \partial_x$ is the constant vector field on $\mathbb{R}$, then the flow on points is given by $\Phi^t: x \mapsto x + t$, and on functions it is given by $(\Phi^t f)(x) = f(x - t)$, so $\partial_x f = -\frac{d}{dt}\Phi^t f$.

Theorem 3.25. *Let G be a finite-dimensional Lie group acting on a manifold M, so we have a map $\rho: G \to \mathrm{Diff}(M)$. Then*

(1) *This action defines a linear map $\rho_*: \mathfrak{g} \to \mathrm{Vect}(M)$.*
(2) *The map ρ_* is a morphism of Lie algebras: $\rho_*[x, y] = [\rho_*(x), \rho_*(y)]$, where the commutator in the right-hand side is the commutator of vector fields.*

If $\mathrm{Diff}(M)$ were a Lie group, this result would be a special case of Proposition 3.12. Since $\mathrm{Diff}(M)$ is not a Lie group, we need to give a separate proof, suitably modifying the proof of Proposition 3.12. We leave this as an exercise to the reader.

We will refer to the map $\rho_*: \mathfrak{g} \to \mathrm{Vect}(M)$ as *action of $\mathfrak{g}$ by vector fields on M*.

Example 3.26. Consider the standard action of $\mathrm{GL}(n, \mathbb{R})$ on $\mathbb{R}^n$. Considering $\mathbb{R}^n$ as a manifold and forgetting the structure of a vector space, we see that each element $a \in \mathfrak{gl}(n, \mathbb{R})$ defines a vector field on $\mathbb{R}^n$. An easy calculation shows that this vector field is given by $v_a(x) = \sum a_{ij}x_j\partial_i$, where $x_1, \ldots x_n$ are the coordinates of a point x in the standard basis of $\mathbb{R}^n$, and $\partial_i = \frac{\partial}{\partial x_i}$.

Another important example is the action of G on itself by left multiplication.

Proposition 3.27. *Consider the action of a Lie group G on itself by left multiplication: $Lg.h = gh$. Then for every $x \in \mathfrak{g}$, the corresponding vector field $\xi = L_*x \in \mathrm{Vect}(G)$ is the* right-invariant *vector field such that $\xi(1) = x$.*

Proof. Consider the one-parameter subgroup $\exp(tx) \subset G$. By Proposition 3.6, for any $g \in G$, we have $L_*x(g) = \frac{d}{dt}|_{t=0}(\exp(tx)g) = xg$. $\qquad\square$

Corollary 3.28. *The isomorphism $\mathfrak{g} \simeq \{$right-invariant vector fields on $G\}$ defined in Theorem 2.27 is an isomorphism of Lie algebras.*

An analog of this statement for left-invariant fields is given in Exercise 3.4.

3.6. Stabilizers and the center

Having developed the basic theory of Lie algebras, we can now go back to proving various results about Lie groups which were announced in Chapter 2, such as proving that the stabilizer of a point is a closed Lie subgroup.

Theorem 3.29. *Let G be a Lie group acting on a manifold M (respectively, a complex Lie group holomorphically acting on a complex manifold M), and let $m \in M$.*

(1) *The stabilizer $G_m = \{g \in G \mid gm = m\}$ is a closed Lie subgroup in G, with Lie algebra $\mathfrak{h} = \{x \in \mathfrak{g} \mid \rho_*(x)(m) = 0\}$, where $\rho_*(x)$ is the vector field on M corresponding to x.*

(2) *The map $G/G_m \to M$ given by $g \mapsto g.m$ is an immersion. Thus, the orbit $\mathcal{O}_m = G \cdot m$ is an immersed submanifold in M, with tangent space $T_m \mathcal{O} = \mathfrak{g}/\mathfrak{h}$.*

Proof. As in the proof of Theorem 2.30, it suffices to show that in some neighborhood U of $1 \in G$ the intersection $U \cap G_m$ is a submanifold with tangent space $T_1 G_m = \mathfrak{h}$.

It easily follows from (3.10) that $\mathfrak{h}$ is closed under commutator, so it is a Lie subalgebra in $\mathfrak{g}$. Also, since for $x \in \mathfrak{h}$, the corresponding vector field $\xi = \rho_*(x)$ vanishes at m, we have $\rho(\exp(tx))(m) = \Phi_\xi^t(m) = m$, so $\exp(tx) \in G_m$.

Now let us choose some vector subspace (not a subalgebra!) $u \subset \mathfrak{g}$ which is complementary to $\mathfrak{h}$: $\mathfrak{g} = \mathfrak{h} \oplus u$. Since the kernel of the map $\rho_*\colon \mathfrak{g} \to T_m M$ is $\mathfrak{h}$, the restriction of this map to u is injective. By implicit function theorem, this implies that the map $u \to M : y \mapsto \rho(\exp(y))(m)$ is injective for sufficiently small $y \in u$, so $\exp(y) \in G_m \iff y = 0$.

Since in a sufficiently small neighborhood U of 1 in G, any element $g \in U$ can be uniquely written in the form $\exp(y)\exp(x), y \in u, x \in \mathfrak{h}$ (which follows from inverse function theorem), and $\exp(y)\exp(x)m = \exp(y)m$, we see that $g \in G_m \iff g \in \exp(\mathfrak{h})$. Since $\exp\mathfrak{h}$ is a submanifold in a neighborhood of $1 \in G$, we see that G_m is a submanifold.

The same proof also shows that we have an isomorphism $T_1(G/G_m) = \mathfrak{g}/\mathfrak{h} \simeq u$, so injectivity of the map $\rho_*\colon u \to T_m M$ shows that the map $G/G_m \to M$ is an immersion. $\square$

This theorem immediately implies a number of corollaries. In particular, we get the following result which was announced in Theorem 2.15.

Corollary 3.30. *Let $f\colon G_1 \to G_2$ be a morphism of real or complex Lie groups, and $f_*\colon \mathfrak{g}_1 \to \mathfrak{g}_2$—the corresponding morphism of Lie algebras. Then $\operatorname{Ker} f$ is a closed Lie subgroup with Lie algebra $\operatorname{Ker} f_*$, and the map $G_1/\operatorname{Ker} f \to G_2$*

is an immersion. If Im f *is a submanifold and thus a closed Lie subgroup, we have a Lie group isomorphism* Im $f \simeq G_1 / \operatorname{Ker} f$.

Proof. Consider the action of G_1 on G_2 given by $\rho(g).h = f(g)h, g \in G_1, h \in G_2$. Then the stabilizer of $1 \in G_2$ is exactly $\operatorname{Ker} f$, so by the previous theorem, it is a closed Lie subgroup with Lie algebra $\operatorname{Ker} f_*$, and $G_1 / \operatorname{Ker} f \hookrightarrow G_2$ is an immersion. $\qquad \square$

Corollary 3.31. *Let V be a representation of a group G, and $v \in V$. Then the stabilizer G_v is a closed Lie subgroup in G with Lie algebra $\{x \in \mathfrak{g} \mid x.v = 0\}$.*

Example 3.32. Let V be a vector space over $\mathbb{K}$ with a bilinear form B, and let

$$\mathrm{O}(V, B) = \{g \in \mathrm{GL}(V) \mid B(g.v, g.w) = B(v, w) \text{ for all } v, w\}$$

be the group of symmetries of B. Then it is a Lie group over $\mathbb{K}$ with the Lie algebra

$$\mathfrak{o}(V, B) = \{x \in \mathfrak{gl}(V) \mid B(x.v, w) + B(v, x.w) = 0 \text{ for all } v, w\}$$

Indeed, define the action of G on the space of bilinear forms by $(gF)(v, w) = F(g^{-1}.v, g^{-1}.w)$. Then $\mathrm{O}(V, B)$ is exactly the stabilizer of B, so by Corollary 3.31, it is a Lie group. Since the corresponding action of $\mathfrak{g}$ is given by $(xF)(v, w) = -F(x.v, w) - F(v, x.w)$ (which follows from Leibniz rule), we get the formula for $\mathfrak{o}(V, B)$.

As special cases, we recover the usual groups $\mathrm{O}(n, \mathbb{K})$ and $\mathrm{Sp}(n, \mathbb{K})$.

Example 3.33. Let A be a finite-dimensional associative algebra over $\mathbb{K}$. Then the group of all automorphisms of A

$$\mathrm{Aut}(A) = \{g \in \mathrm{GL}(A) \mid (ga) \cdot (gb) = g(a \cdot b) \text{ for all } a, b \in A\}$$

is a Lie group with Lie algebra

$$\mathrm{Der}(A) = \{x \in \mathfrak{gl}(A) \mid (x.a)b + a(x.b) = x.(ab) \text{ for all } a, b \in A\} \qquad (3.12)$$

(this Lie algebra is called the algebra of *derivations* of A).

Indeed, if we consider the space W of all linear maps $A \otimes A \to A$ and define the action of G by $(g.f)(a \otimes b) = gf(g^{-1}a \otimes g^{-1}b)$ then $\mathrm{Aut} A = G_\mu$, where $\mu : A \otimes A \to A$ is the multiplication. So by Corollary 3.31, $\mathrm{Aut}(A)$ is a Lie group with Lie algebra $\mathrm{Der}(A)$.

The same argument also shows that for a finite-dimensional Lie algebra $\mathfrak{g}$, the group

$$\mathrm{Aut}(\mathfrak{g}) = \{g \in \mathrm{GL}(\mathfrak{g}) \mid [ga, gb] = g[a, b] \text{ for all } a, b \in \mathfrak{g}\} \tag{3.13}$$

is a Lie group with Lie algebra

$$\mathrm{Der}(\mathfrak{g}) = \{x \in \mathfrak{gl}(\mathfrak{g}) \mid [x.a, b] + [a, x.b] = x.[a, b] \text{ for all } a, b \in \mathfrak{g}\} \tag{3.14}$$

called the Lie algebra of derivations of $\mathfrak{g}$. This algebra will play an important role in the future.

Finally, we can show that the center of G is a closed Lie subgroup.

Definition 3.34. Let $\mathfrak{g}$ be a Lie algebra. The center of $\mathfrak{g}$ is defined by

$$\mathfrak{z}(\mathfrak{g}) = \{x \in \mathfrak{g} \mid [x, y] = 0 \; \forall y \in \mathfrak{g}\}.$$

Obviously, $\mathfrak{z}(\mathfrak{g})$ is an ideal in $\mathfrak{g}$.

Theorem 3.35. *Let G be a connected Lie group. Then its center $Z(G)$ is a closed Lie subgroup with Lie algebra $\mathfrak{z}(\mathfrak{g})$. If G is not connected, then $Z(G)$ is still a closed Lie subgroup; however, its Lie algebra might be smaller than $\mathfrak{z}(\mathfrak{g})$.*

Proof. Let $g \in G$, $x \in \mathfrak{g}$. It follows from the identity $\exp(\mathrm{Ad}\, g.tx) = g \exp(tx) g^{-1}$ that g commutes with all elements of one-parameter subgroup $\exp(tx)$ iff $\mathrm{Ad}\, g.x = x$. Since for a connected Lie group, elements of the form $\exp(tx)$ generate G, we see that $g \in Z(G) \iff \mathrm{Ad}\, g.x = x$ for all $x \in \mathfrak{g}$. In other words, $Z(G) = \mathrm{Ker}\,\mathrm{Ad}$, where $\mathrm{Ad} \colon G \to \mathrm{GL}(\mathfrak{g})$ is given by the adjoint action. Now the result follows from Corollary 3.30. $\qquad\square$

The quotient group $G/Z(G)$ is usually called the *adjoint group* associated with G and denoted $\mathrm{Ad}\, G$:

$$\mathrm{Ad}\, G = G/Z(G) = \mathrm{Im}(\mathrm{Ad} \colon G \to \mathrm{GL}(\mathfrak{g})) \tag{3.15}$$

(for connected G). The corresponding Lie algebra is

$$\mathrm{ad}\, \mathfrak{g} = \mathfrak{g}/\mathfrak{z}(\mathfrak{g}) = \mathrm{Im}(\mathrm{ad} \colon \mathfrak{g} \to \mathfrak{gl}(\mathfrak{g})). \tag{3.16}$$

3.7. Campbell–Hausdorff formula

So far, we have shown that the multiplication in a Lie group G defines the commutator in $\mathfrak{g} = T_1 G$. However, the definition of commutator (3.2) only used the lowest non-trivial term of the group law in logarithmic coordinates. Thus, it might be expected that higher terms give more operations on $\mathfrak{g}$. However, it turns out that it is not so: the whole group law is completely determined by the lowest term, i.e. by the commutator. The following theorem gives the first indication of this.

Theorem 3.36. *Let* $x, y \in \mathfrak{g}$ *be such that* $[x, y] = 0$. *Then* $\exp(x) \exp(y) = \exp(x + y) = \exp(y) \exp(x)$.

Proof. The most instructive (but not the easiest; see Exercise 3.12) way of deducing this theorem is as follows. Let ξ, η be right-invariant vector fields corresponding to x, y respectively, and let Φ_ξ^t, Φ_η^t be time t flows of these vector fields respectively (see Section 3.5). By Corollary 3.28, $[\xi, \eta] = 0$. By (3.9), it implies that $\frac{d}{dt}(\Phi_\xi^t)_* \eta = 0$, which implies that $(\Phi_\xi^t)_* \eta = \eta$, i.e. the flow of ξ preserves field η. This, in turn, implies that Φ_ξ^t commutes with the flow of field η, so $\Phi_\xi^t \Phi_\eta^s \Phi_\xi^{-t} = \Phi_\eta^s$. Applying this to point $1 \in G$ and using Proposition 3.6, we get $\exp(tx) \exp(sy) \exp(-tx) = \exp(sy)$, so $\exp(tx), \exp(sy)$ commute for all values of s, t.

In particular, this implies that $\exp(tx) \exp(ty)$ is a one-parameter subgroup; computing the tangent vector at $t = 0$, we see that $\exp(tx) \exp(ty) = \exp(t(x + y))$. $\square$

In fact, similar ideas allow one to prove the following general statement, known as the Campbell–Hausdorff formula.

Theorem 3.37. *For small enough* $x, y \in \mathfrak{g}$ *one has*

$$\exp(x) \exp(y) = \exp(\mu(x, y))$$

for some $\mathfrak{g}$-*valued function* $\mu(x, y)$ *which is given by the following series convergent in some neighborhood of* $(0, 0)$:

$$\mu(x, y) = x + y + \sum_{n \geq 2} \mu_n(x, y), \tag{3.17}$$

where $\mu_n(x, y)$ *is a Lie polynomial in* x, y *of degree* n, *i.e. an expression consisting of commutators of* x, y, *their commutators, etc., of total degree* n *in* x, y. *This expression is universal: it does not depend on the Lie algebra* $\mathfrak{g}$ *or on the choice of* x, y.

It is possible to write the expression for μ explicitly (see, e.g., [10]). However, this is rarely useful, so we will only write the first few terms:

$$\mu(x, y) = x + y + \frac{1}{2}[x, y] + \frac{1}{12}\big([x, [x, y]] + [y, [y, x]]\big) + \cdots \tag{3.18}$$

The proof of this theorem is rather long. The key idea is writing the differential equation for the function $Z(t) = \mu(tx, y)$; the right-hand side of this equation will be a power series of the form $\sum a_n t^n (\operatorname{ad} x)^n y$. Solving this differential equation by power series gives the Campbell–Hausdorff formula. Details of the proof can be found, for example, in [10, Section 1.6].

Corollary 3.38. *The group operation in a connected Lie group G can be recovered from the commutator in $\mathfrak{g} = T_1 G$.*

Indeed, locally the group law is determined by the Campbell–Hausdorff formula, and G is generated by a neighborhood of 1.

Note, however, that by itself this corollary does not allow us to recover the group G from its Lie algebra $\mathfrak{g}$: it only allows us to determine the group law provided that we already know the structure of G as a manifold.

3.8. Fundamental theorems of Lie theory

Let us summarize the results we have so far about the relation between Lie groups and Lie algebras.

(1) Every real or complex Lie group G defines a Lie algebra $\mathfrak{g} = T_1 G$ (respectively, real or complex), with commutator defined by (3.2); we will write $\mathfrak{g} = \operatorname{Lie}(G)$. Every morphism of Lie groups $\varphi \colon G_1 \to G_2$ defines a morphism of Lie algebras $\varphi_* \colon \mathfrak{g}_1 \to \mathfrak{g}_2$. For connected G_1, the map

$$\operatorname{Hom}(G_1, G_2) \to \operatorname{Hom}(\mathfrak{g}_1, \mathfrak{g}_2)$$

$$\varphi \mapsto \varphi_*$$

is injective. (Here $\operatorname{Hom}(\mathfrak{g}_1, \mathfrak{g}_2)$ is the set of Lie algebra morphisms.)
(2) As a special case of the previous, every Lie subgroup $H \subset G$ defines a Lie subalgebra $\mathfrak{h} \subset \mathfrak{g}$.
(3) The group law in a connected Lie group G can be recovered from the commutator in $\mathfrak{g}$; however, we do not yet know whether we can also recover the topology of G from $\mathfrak{g}$.

However, this still leaves a number of questions.

(1) Given a morphism of Lie algebras $\mathfrak{g}_1 \to \mathfrak{g}_2$, where $\mathfrak{g}_1 = \operatorname{Lie}(G_1), \mathfrak{g}_2 = \operatorname{Lie}(G_2)$, can this morphism be always lifted to a morphism of the Lie groups?

(2) Given a Lie subalgebra $\mathfrak{h} \subset \mathfrak{g} = \mathrm{Lie}(G)$, does there always exist a corresponding Lie subgroup $H \subset G$?
(3) Can every Lie algebra be obtained as a Lie algebra of a Lie group?

As the following example shows, in this form the answer to question 1 is negative.

Example 3.39. Let $G_1 = S^1 = \mathbb{R}/\mathbb{Z}, G_2 = \mathbb{R}$. Then the Lie algebras are $\mathfrak{g}_1 = \mathfrak{g}_2 = \mathbb{R}$ with zero commutator. Consider the identity map $\mathfrak{g}_1 \to \mathfrak{g}_2 : a \mapsto a$. Then the corresponding morphism of Lie groups, if it exists, should be given by $\theta \mapsto \theta$; on the other hand, it must also satisfy $f(\mathbb{Z}) = \{0\}$. Thus, this morphism of Lie algebras can not be lifted to a morphism of Lie groups.

In this example the difficulty arose because G_1 was not simply-connected. It turns out that this is the only difficulty: after taking care of this, the answers to all the questions posed above are positive. The following theorems give precise statements.

Theorem 3.40. *For any real or complex Lie group G, there is a bijection between connected Lie subgroups $H \subset G$ and Lie subalgebras $\mathfrak{h} \subset \mathfrak{g}$, given by $H \to \mathfrak{h} = \mathrm{Lie}(H) = T_1 H$.*

Theorem 3.41. *If G_1, G_2 are Lie groups (real or complex) and G_1 is connected and simply connected, then $\mathrm{Hom}(G_1, G_2) = \mathrm{Hom}(\mathfrak{g}_1, \mathfrak{g}_2)$, where $\mathfrak{g}_1, \mathfrak{g}_2$ are Lie algebras of G_1, G_2 respectively.*

Theorem 3.42 (Lie's third theorem). *Any finite-dimensional real or complex Lie algebra is isomorphic to a Lie algebra of a Lie group (respectively, real or complex).*

Theorems 3.40–3.42 are the fundamental theorems of Lie theory; their proofs are discussed below. In particular, combining these theorems with the previous results, we get the following important corollary.

Corollary 3.43. *For any real or complex finite-dimensional Lie algebra $\mathfrak{g}$, there is a unique (up to isomorphism) connected simply-connected Lie group G (respectively, real or complex) with $\mathrm{Lie}(G) = \mathfrak{g}$. Any other connected Lie group G' with Lie algebra $\mathfrak{g}$ must be of the form G/Z for some discrete central subgroup $Z \subset G$.*

Proof. By Theorem 3.42, there is a Lie group with Lie algebra $\mathfrak{g}$. Taking the universal cover of the connected component of identity in this group (see Theorem 2.7), we see that there exists a connected, simply-connected G with $\mathrm{Lie}(G) = \mathfrak{g}$. By Theorem 3.41, if G' is another connected Lie group with

Lie algebra $\mathfrak{g}$, then there is a group homomorphism $G \to G'$ which is locally an isomorphism; thus, by results of Exercise 2.3, $G' = G/Z$ for some discrete central subgroup Z.

Uniqueness of simply-connected group G now follows from $\pi_1(G/Z) = Z$ (Theorem 2.7). □

This corollary can be reformulated as follows.

Corollary 3.44. *The categories of finite-dimensional Lie algebras and connected, simply-connected Lie groups are equivalent.*

We now turn to the discussion of the proofs of the fundamental theorems.

Proof of Theorem 3.42. The proof of this theorem is rather complicated and full details will not be given here. The basic idea is to show that any Lie algebra is isomorphic to a subalgebra in $\mathfrak{gl}(n, \mathbb{K})$ (this statement is known as the Ado theorem), after which we can use Theorem 3.40. However, the proof of the Ado theorem is long and requires a lot of structure theory of Lie algebras, some of which will be given in the subsequent chapters. The simplest case is when the Lie algebra has no center (that is, $\operatorname{ad} x \neq 0$ for all x), then $x \mapsto \operatorname{ad} x$ gives an embedding $\mathfrak{g} \subset \mathfrak{gl}(\mathfrak{g})$. Proof of the general case can be found, e.g., in [24]. □

Proof of Theorem 3.41. We will show that this theorem follows from Theorem 3.40. Indeed, we already discussed that any morphism of Lie groups defines a morphism of Lie algebras and that for connected G_1, the map $\operatorname{Hom}(G_1, G_2) \to \operatorname{Hom}(\mathfrak{g}_1, \mathfrak{g}_2)$ is injective (see Theorem 3.20). Thus, it remains to show that it is surjective, i.e. that every morphism of Lie algebras $f : \mathfrak{g}_1 \to \mathfrak{g}_2$ can be lifted to a morphism of Lie groups $\varphi : G_1 \to G_2$ with $\varphi_* = f$.

Define $G = G_1 \times G_2$. Then the Lie algebra of G is $\mathfrak{g}_1 \times \mathfrak{g}_2$. Let $\mathfrak{h} = \{(x, f(x)) \mid x \in \mathfrak{g}_1\} \subset \mathfrak{g}$. This is a subalgebra: it is obviously a subspace, and $[(x, f(x)), (y, f(y))] = ([x, y], [f(x), f(y)]) = ([x, y], f([x, y]))$ (the last identity uses that f is a morphism of Lie algebras). Theorem 3.40 implies that there is a corresponding connected Lie subgroup $H \hookrightarrow G_1 \times G_2$. Composing this embedding with the projection $p : G_1 \times G_2 \to G_1$, we get a morphism of Lie groups $\pi : H \to G_1$, and $\pi_* : \mathfrak{h} = \operatorname{Lie}(H) \to \mathfrak{g}_1$ is an isomorphism. By results of Exercise 2.3, π is a covering map. On the other hand, G_1 is simply-connected, and H is connected, so π must be an isomorphism. Thus, we have an inverse map $\pi^{-1} : G_1 \to H$.

Now construct the map $\varphi : G_1 \to G_2$ as a composition $G_1 \xrightarrow{\pi^{-1}} H \hookrightarrow G_1 \times G_2 \to G_2$. By definition, it is a morphism of Lie groups, and $\varphi_* : \mathfrak{g}_1 \to \mathfrak{g}_2$ is the composition $x \mapsto (x, f(x)) \mapsto f(x)$. Thus, we have lifted f to a morphism of Lie groups. □

Remark 3.45. In fact, the arguments above can be reversed to deduce Theorem 3.40 from Theorem 3.41. For example, this is the way these theorems are proved in [41].

Proof of Theorem 3.40. We will give the proof in the real case; proof in the complex case is similar.

The proof we give here is based on the notion of integrable distribution. For the reader's convenience, we give the basic definitions here; details can be found in [49] or [55].

A k-dimensional *distribution* on a manifold M is a k-dimensional subbundle $\mathcal{D} \subset TM$. In other words, at every point $p \in M$ we have a k-dimensional subspace $\mathcal{D}_p \subset T_pM$, which smoothly depends on p. This is a generalization of a well-known notion of direction field, commonly used in the theory of ordinary differential equations. For a vector field v we will write $v \in \mathcal{D}$ if for every point p we have $v(p) \in \mathcal{D}_p$.

An integral manifold for a distribution $\mathcal{D}$ is a k-dimensional submanifold $X \subset M$ such that at every point $p \in X$, we have $T_pX = \mathcal{D}_p$. Again, this is a straightforward generalization of the notion of an integral curve. However, for $k > 1$, existence of integral manifolds (even locally) is not automatic. We say that a distribution $\mathcal{D}$ is *completely integrable* if for every $p \in M$, locally there exists an integral manifold containing p (it is easy to show that such an integral manifold is unique).

The following theorem gives a necessary and sufficient criterion of integrability of a distribution.

Theorem 3.46 (Frobenius integrability criterion). *A distribution $\mathcal{D}$ on M is completely integrable if and only if for any two vector fields $\xi, \eta \in \mathcal{D}$, one has $[\xi, \eta] \in \mathcal{D}$.*

Proof of this theorem can be found in many books on differential geometry, such as [49] and [55], and will not be repeated here.

Integrability is a local condition: it guarantees existence of an integral manifold in a neighborhood of a given point. It is natural to ask whether this local integral manifold can be extended to give a closed submanifold of M. The following theorem gives the answer.

Theorem 3.47. *Let $\mathcal{D}$ be a completely integrable distribution on M. Then for every point $p \in M$, there exists a unique connected immersed integral submanifold $N \subset M$ of D which contains p and is maximal, i.e. contains any other connected immersed integral submanifold containing p.*

Note, however, that the integral submanifold needs not be closed: in general, it is not even an embedded submanifold but only an immersed one.

As before, we refer the reader to [49], [55] for the proof.

Let us now apply this theory to constructing, for a given Lie group G and subalgebra $\mathfrak{h} \subset \mathfrak{g}$, the corresponding Lie subgroup $H \subset G$.

Notice that if such an H exists, then at every point $p \in H$, $T_p H = (T_1 H)p = \mathfrak{h}.p$ (as in Section 2.6, we use notation $v.p$ as a shorthand for $(R_{p^{-1}})_* v$). Thus, H will be an integral manifold of the distribution $\mathcal{D}^{\mathfrak{h}}$ defined by $\mathcal{D}_p^{\mathfrak{h}} = \mathfrak{h}.p$. Let us use this to construct H.

Lemma 3.48. *For every point $g \in G$, there is locally an integral manifold of the distribution $\mathcal{D}^{\mathfrak{h}}$ containing g, namely $H^0 \cdot g$, where $H^0 = \exp u$ for some neighborhood u of 0 in $\mathfrak{h}$.*

This lemma can be easily proved using Frobenius theorem. Indeed, the distribution $\mathcal{D}^{\mathfrak{h}}$ is generated by right-invariant vector fields corresponding to elements of $\mathfrak{h}$. Since $\mathfrak{h}$ is closed under $[,]$, and commutator of right invariant vector fields coincides with the commutator in $\mathfrak{g}$ (Corollary 3.28), this shows that the space of fields tangent to $\mathcal{D}^{\mathfrak{h}}$ is closed under the commutator, and thus $\mathcal{D}^{\mathfrak{h}}$ is completely integrable.

To get an explicit description of the integral manifold, note that by Proposition 3.6, the curve $e^{tx}g$ for $x \in \mathfrak{h}$ is the integral curve for a right invariant vector field corresponding to x and thus this curve must be in the integral manifold. Thus, for small enough $x \in \mathfrak{h}$, $\exp(x)g$ is in the integral manifold passing through g. Comparing dimensions we get the statement of the lemma.

Alternatively, this lemma can also be proved without use of Frobenius theorem but using the Campbell–Hausdorff formula instead.

Now that we have proved the lemma, we can construct the immersed subgroup H as the maximal connected immersed integral manifold containing 1 (see Theorem 3.47). The only thing which remains to be shown is that H is a subgroup. But since the distribution $\mathcal{D}^{\mathfrak{h}}$ is right-invariant, right action of G on itself sends integral manifolds to integral manifolds; therefore, for any $p \in H$, $H \cdot p$ will be an integral manifold for $\mathcal{D}^{\mathfrak{h}}$ containing p. Since H itself also contains p, we must have $H \cdot p = H$, so H is a subgroup. $\square$

3.9. Complex and real forms

An interesting application of the correspondence between Lie groups and Lie algebras is the interplay between real and complex Lie algebras and groups.

Definition 3.49. Let $\mathfrak{g}$ be a real Lie algebra. Its *complexification* is the complex Lie algebra $\mathfrak{g}_{\mathbb{C}} = \mathfrak{g} \otimes_{\mathbb{R}} \mathbb{C} = \mathfrak{g} \oplus i\mathfrak{g}$ with the obvious commutator. In this situation, we will also say that $\mathfrak{g}$ is a real form of $\mathfrak{g}_{\mathbb{C}}$.

In some cases, complexification is obvious: for example, if $\mathfrak{g} = \mathfrak{sl}(n, \mathbb{R})$, then $\mathfrak{g}_{\mathbb{C}} = \mathfrak{sl}(n, \mathbb{C})$. The following important example, however, is less obvious.

Example 3.50. Let $\mathfrak{g} = \mathfrak{u}(n)$. Then $\mathfrak{g}_{\mathbb{C}} = \mathfrak{gl}(n, \mathbb{C})$.

Indeed, this immediately follows from the fact that any complex matrix can be uniquely written as a sum of skew-Hermitian (i.e., from $\mathfrak{u}(n)$) and Hermitian $(i\mathfrak{u}(n))$ matrices.

These notions can be extended to Lie groups. For simplicity, we only consider the case of connected groups.

Definition 3.51. Let G be a connected complex Lie group, $\mathfrak{g} = \mathrm{Lie}(G)$ and let $K \subset G$ be a closed real Lie subgroup in G such that $\mathfrak{k} = \mathrm{Lie}(K)$ is a real form of $\mathfrak{g}$. Then K is called a *real form* of G.

It can be shown (see Exercise 3.15) that if $\mathfrak{g} = \mathrm{Lie}(G)$ is the Lie algebra of a connected simply-connected complex Lie group G, then every real form $\mathfrak{k} \subset \mathfrak{g}$ can be obtained from a real form $K \subset G$ of the Lie group.

Going in the opposite direction, from a real Lie group to a complex one, is more subtle: there are real Lie groups that can not be obtained as real forms of a complex Lie group (for example, it is known that the universal cover of $\mathrm{SL}(2, \mathbb{R})$ is not a real form of any complex Lie group). It is still possible to define a complexification $G_{\mathbb{C}}$ for any real Lie group G; however, in general G is not a subgroup of $G_{\mathbb{C}}$. Detailed discussion of this can be found in [15, Section I.7].

Example 3.52. The group $G = \mathrm{SU}(n)$ is a compact real form of the complex group $\mathrm{SL}(n, \mathbb{C})$.

The operation of complexification, which is trivial at the level of Lie algebras, is highly non-trivial at the level of Lie groups. Lie groups G and $G_{\mathbb{C}}$ may be topologically quite different: for example, $\mathrm{SU}(n)$ is compact while $\mathrm{SL}(n, \mathbb{C})$ is not. On the other hand, it is natural to expect – and is indeed so, as we will show later – that $\mathfrak{g}$ and $\mathfrak{g}_{\mathbb{C}}$ share many algebraic properties, such as semisimplicity. Thus, we may use, for example, compact group $\mathrm{SU}(n)$ to prove some results about the non-compact group $\mathrm{SL}(n, \mathbb{C})$. Moreover, since $\mathfrak{sl}(n, \mathbb{R})_{\mathbb{C}} = \mathfrak{sl}(n, \mathbb{C})$, this will also give us results about the non-compact real group $\mathrm{SL}(n, \mathbb{R})$. We will give an application of this to the study of representations of Lie groups in the next chapter.

3.10. Example: $\mathfrak{so}(3, \mathbb{R})$, $\mathfrak{su}(2)$, and $\mathfrak{sl}(2, \mathbb{C})$

In this section, we bring together various explicit formulas related to Lie algebras $\mathfrak{so}(3, \mathbb{R})$, $\mathfrak{su}(2)$, $\mathfrak{sl}(2, \mathbb{C})$. Most of these results have appeared before in various examples and exercises; this section brings them together for the reader's convenience. This section contains no proofs: they are left to the reader as exercises.

3.10.1. Basis and commutation relations

A basis in $\mathfrak{so}(3, \mathbb{R})$ is given by

$$J_x = \begin{pmatrix} 0 & 0 & 0 \\ 0 & 0 & -1 \\ 0 & 1 & 0 \end{pmatrix}, \qquad J_y = \begin{pmatrix} 0 & 0 & 1 \\ 0 & 0 & 0 \\ -1 & 0 & 0 \end{pmatrix}, \qquad J_z = \begin{pmatrix} 0 & -1 & 0 \\ 1 & 0 & 0 \\ 0 & 0 & 0 \end{pmatrix} \tag{3.19}$$

The corresponding one-parameter subgroups in $SO(3, \mathbb{R})$ are rotations: $\exp(tJ_x)$ is rotation by angle t around x-axis, and similarly for y, z.

The commutation relations are given by

$$[J_x, J_y] = J_z, \qquad [J_y, J_z] = J_x, \qquad [J_z, J_x] = J_y. \tag{3.20}$$

A basis in $\mathfrak{su}(2)$ is given by so-called Pauli matrices multiplied by i:

$$i\sigma_1 = \begin{pmatrix} 0 & i \\ i & 0 \end{pmatrix}, \qquad i\sigma_2 = \begin{pmatrix} 0 & 1 \\ -1 & 0 \end{pmatrix}, \qquad i\sigma_3 = \begin{pmatrix} i & 0 \\ 0 & -i \end{pmatrix}. \tag{3.21}$$

The commutation relations are given by

$$[i\sigma_1, i\sigma_2] = -2i\sigma_3, \qquad [i\sigma_2, i\sigma_3] = -2i\sigma_1, \qquad [i\sigma_3, i\sigma_1] = -2i\sigma_2. \tag{3.22}$$

Since $\mathfrak{sl}(2, \mathbb{C}) = \mathfrak{su}(2) \otimes \mathbb{C}$, the same matrices can also be taken as a basis of $\mathfrak{sl}(2, \mathbb{C})$. However, it is customary to use the following basis in $\mathfrak{sl}(2, \mathbb{C})$:

$$e = \begin{pmatrix} 0 & 1 \\ 0 & 0 \end{pmatrix}, \qquad f = \begin{pmatrix} 0 & 0 \\ 1 & 0 \end{pmatrix}, \qquad h = \begin{pmatrix} 1 & 0 \\ 0 & -1 \end{pmatrix}. \tag{3.23}$$

In this basis, the commutation relations are given by

$$[e, f] = h, \qquad [h, e] = 2e, \qquad [h, f] = -2f. \tag{3.24}$$

3.10.2. Invariant bilinear form

Each of these Lie algebras has an Ad G-invariant symmetric bilinear form. In each case, it can be defined by $(x,y) = -\text{tr}(xy)$ (of course, it could also be defined without the minus sign). For $\mathfrak{so}(3,\mathbb{R})$, this form can also be rewritten as $(x,y) = \text{tr}(xy^t)$; for $\mathfrak{su}(n)$, as $(x,y) = \text{tr}(x\bar{y}^t)$, which shows that in these two cases this form is positive definite. In terms of bases defined above, it can be written as follows:

- $\mathfrak{so}(3,\mathbb{R})$: elements J_x, J_y, J_z are orthogonal to each other, and $(J_x, J_x) = (J_y, J_y) = (J_z, J_z) = 2$
- $\mathfrak{su}(2)$: elements $i\sigma_k$ are orthogonal, and $(i\sigma_k, i\sigma_k) = 2$.
- $\mathfrak{sl}(2,\mathbb{C})$: $(h,h) = -2$, $(e,f) = (f,e) = -1$, all other products are zero.

3.10.3. Isomorphisms

We have an isomorphism of Lie algebras $\mathfrak{su}(2) \xrightarrow{\sim} \mathfrak{so}(3,\mathbb{R})$ given by

$$
\begin{aligned}
i\sigma_1 &\mapsto -2J_x \\
i\sigma_2 &\mapsto -2J_y \\
i\sigma_3 &\mapsto -2J_z.
\end{aligned}
\tag{3.25}
$$

It can be lifted to a morphism of Lie groups $\text{SU}(2) \to \text{SO}(3,\mathbb{R})$, which is a twofold cover (see Exercise 2.8).

The inclusion $\mathfrak{su}(2) \subset \mathfrak{sl}(2,\mathbb{C})$ gives an isomorphism $\mathfrak{su}(2)_{\mathbb{C}} \simeq \mathfrak{sl}(2,\mathbb{C})$. In terms of basis, it is given by

$$
\begin{aligned}
i\sigma_1 &\mapsto i(e+f) \\
i\sigma_2 &\mapsto e-f \\
i\sigma_3 &\mapsto ih.
\end{aligned}
\tag{3.26}
$$

Combining these two isomorphisms, we get an isomorphism $\mathfrak{so}(3,\mathbb{R})_{\mathbb{C}} = \mathfrak{so}(3,\mathbb{C}) \xrightarrow{\sim} \mathfrak{sl}(2,\mathbb{C})$

$$
\begin{aligned}
J_x &\mapsto -\frac{i}{2}(e+f) \\
J_y &\mapsto \frac{1}{2}(f-e) \\
J_z &\mapsto -\frac{ih}{2}.
\end{aligned}
\tag{3.27}
$$

3.11. Exercises

3.1. Consider the group $SL(2, \mathbb{R})$. Show that the element $X = \begin{pmatrix} -1 & 1 \\ 0 & -1 \end{pmatrix}$ is not in the image of the exponential map. (Hint: if $X = \exp(x)$, what are the eigenvalues of x?).

3.2. Let $f \colon \mathfrak{g} \to G$ be any smooth map such that $f(0) = 1, f_*(0) = \mathrm{id}$; we can view such a map as a local coordinate system near $1 \in G$. Show that the the group law written in this coordinate system has the form $f(x)f(y) = f(x + y + B(x, y) + \cdots)$ for some bilinear map $B \colon \mathfrak{g} \otimes \mathfrak{g} \to \mathfrak{g}$ and that $B(x, y) - B(y, x) = [x, y]$.

3.3. Show that all forms of Jacobi identity given in (3.5), (3.6) are equivalent.

3.4. Show that if we denote, for $x \in \mathfrak{g}$, by ξ_x the *left-invariant* vector field on G such that $\xi_x(1) = x$ (cf. Theorem 2.27), then $[\xi_x, \xi_y] = -\xi_{[x,y]}$.

3.5. (1) Prove that $\mathbb{R}^3$ with the commutator given by the cross-product is a Lie algebra. Show that this Lie algebra is isomorphic to $\mathfrak{so}(3, \mathbb{R})$.

(2) Let $\varphi \colon \mathfrak{so}(3, \mathbb{R}) \to \mathbb{R}^3$ be the isomorphism of part (1). Prove that under this isomorphism, the standard action of $\mathfrak{so}(3)$ on $\mathbb{R}^3$ is identified with the action of $\mathbb{R}^3$ on itself given by the cross-product:

$$ a \cdot \vec{v} = \varphi(a) \times \vec{v}, \qquad a \in \mathfrak{so}(3), \vec{v} \in \mathbb{R}^3 $$

where $a \cdot \vec{v}$ is the usual multiplication of a matrix by a vector.

This problem explains a common use of cross-products in mechanics (see, e.g. [1]): angular velocities and angular momenta are actually elements of Lie algebra $\mathfrak{so}(3, \mathbb{R})$ (to be precise, angular momenta are elements of the dual vector space, $(\mathfrak{so}(3, \mathbb{R})^*)$, but we can ignore this difference). To avoid explaining this, most textbooks write angular velocities as vectors in $\mathbb{R}^3$ and use cross-product instead of commutator. Of course, this would completely fail in dimensions other than 3, where $\mathfrak{so}(n, \mathbb{R})$ is not isomorphic to $\mathbb{R}^n$ even as a vector space.

3.6. Let P_n be the space of polynomials with real coefficients of degree $\leq n$ in variable x. The Lie group $G = \mathbb{R}$ acts on P_n by translations of the argument: $\rho(t)(x) = x + t, t \in G$. Show that the corresponding action of the Lie algebra $\mathfrak{g} = \mathbb{R}$ is given by $\rho(a) = a\partial_x, a \in \mathfrak{g}$, and deduce from this the Taylor formula for polynomials:

$$ f(x + t) = \sum_{n \geq 0} \frac{(t\partial_x)^n}{n!} f. $$

3.7. Let G be the Lie group of all maps $A \colon \mathbb{R} \to \mathbb{R}$ having the form $A(x) = ax + b, a \neq 0$. Describe explicitly the corresponding Lie algebra. [There are two ways to do this problem. The easy way is to embed $G \subset GL(2, \mathbb{R})$, which makes the problem trivial. A more straightforward way is to explicitly construct some basis in the tangent space, construct the corresponding one-parameter subgroups, and compute the commutator using (3.3). The second way is recommended to those who want to understand how the correspondence between Lie groups and Lie algebras works.]

3.8. Let $SL(2, \mathbb{C})$ act on $\mathbb{CP}^1$ in the usual way:

$$\begin{bmatrix} a & b \\ c & d \end{bmatrix} (x : y) = (ax + by : cx + dy).$$

This defines an action of $\mathfrak{g} = \mathfrak{sl}(2, \mathbb{C})$ by vector fields on $\mathbb{CP}^1$. Write explicitly vector fields corresponding to h, e, f in terms of coordinate $t = x/y$ on the open cell $\mathbb{C} \subset \mathbb{CP}^1$.

3.9. Let G be a Lie group with Lie algebra $\mathfrak{g}$, and $\mathrm{Aut}(\mathfrak{g})$, $\mathrm{Der}(\mathfrak{g})$ be as defined in Example 3.33.
  (1) Show that $g \mapsto \mathrm{Ad}\, g$ gives a morphism of Lie groups $G \to \mathrm{Aut}(\mathfrak{g})$; similarly, $x \mapsto \mathrm{ad}\, x$ is a morphism of Lie algebras $\mathfrak{g} \to \mathrm{Der}\, \mathfrak{g}$. (The automorphisms of the form $\mathrm{Ad}\, g$ are called *inner* automorphisms; the derivations of the form $\mathrm{ad}\, x, x \in \mathfrak{g}$ are called inner derivations.)
  (2) Show that for $f \in \mathrm{Der}\, \mathfrak{g}$, $x \in \mathfrak{g}$, one has $[f, \mathrm{ad}\, x] = \mathrm{ad}\, f(x)$ as operators in $\mathfrak{g}$, and deduce from this that $\mathrm{ad}(\mathfrak{g})$ is an ideal in $\mathrm{Der}\, \mathfrak{g}$.

3.10. Let $\{H_\alpha\}_{\alpha \in A}$ be some family of closed Lie subgroups in G, with the Lie algebras $\mathfrak{h}_\alpha = \mathrm{Lie}(H_\alpha)$. Let $H = \bigcap_\alpha H_\alpha$. Without using the theorem about closed subgroup, show that H is a closed Lie subgroup with Lie algebra $\mathfrak{h} = \bigcap_\alpha \mathfrak{h}_\alpha$.

3.11. Let J_x, J_y, J_z be the basis in $\mathfrak{so}(3, \mathbb{R})$ described in Section 3.10. The standard action of $SO(3, \mathbb{R})$ on $\mathbb{R}^3$ defines an action of $\mathfrak{so}(3, \mathbb{R})$ by vector fields on $\mathbb{R}^3$. Abusing the language, we will use the same notation J_x, J_y, J_z for the corresponding vector fields on $\mathbb{R}^3$. Let $\Delta_{\mathrm{sph}} = J_x^2 + J_y^2 + J_z^2$; this is a second order differential operator on $\mathbb{R}^3$, which is usually called the *spherical Laplace operator*, or the *Laplace operator on the sphere*.
  (1) Write Δ_{sph} in terms of $x, y, z, \partial_x, \partial_y, \partial_z$.
  (2) Show that Δ_{sph} is well defined as a differential operator on a sphere $S^2 = \{(x, y, z) \mid x^2 + y^2 + z^2 = 1\}$, i.e., if f is a function on $\mathbb{R}^3$ then $(\Delta_{\mathrm{sph}} f)|_{S^2}$ only depends on $f|_{S^2}$.

(3) Show that the usual Laplace operator $\Delta = \partial_x^2 + \partial_y^2 + \partial_z^2$ can be written in the form $\Delta = \frac{1}{r^2}\Delta_{sph} + \Delta_{radial}$, where Δ_{radial} is a differential operator written in terms of $r = \sqrt{x^2 + y^2 + z^2}$ and $r\partial_r = x\partial_x + y\partial_y + z\partial_z$.

(4) Show that Δ_{sph} is rotation invariant: for any function f and $g \in SO(3, \mathbb{R})$, $\Delta_{sph}(gf) = g(\Delta_{sph}f)$. (Later we will describe a better way of doing this.)

3.12. Give an alternative proof of Theorem 3.36, using Lemma 3.15.

3.13. (1) Let $\mathfrak{g}$ be a three-dimensional real Lie algebra with basis x, y, z and commutation relations $[x, y] = z$, $[z, x] = [z, y] = 0$ (this algebra is called *Heisenberg algebra*). Without using the Campbell–Hausdorff formula, show that in the corresponding Lie group, one has $\exp(tx)\exp(sy) = \exp(tsz)\exp(sy)\exp(tx)$ and construct explicitly the connected, simply connected Lie group corresponding to $\mathfrak{g}$.

(2) Generalize the previous part to the Lie algebra $\mathfrak{g} = V \oplus \mathbb{R}z$, where V is a real vector space with non-degenerate skew-symmetric form ω and the commutation relations are given by $[v_1, v_2] = \omega(v_1, v_2)z$, $[z, v] = 0$.

3.14. This problem is for readers familiar with the mathematical formalism of classical mechanics.

Let G be a real Lie group and A – a positive definite symmetric bilinear form on $\mathfrak{g}$; such a form can also be considered as a linear map $\mathfrak{g} \to \mathfrak{g}^*$.

(1) Let us extend A to a left invariant metric on G. Consider mechanical system describing free motion of a particle on G, with kinetic energy given by $A(\dot{g}, \dot{g})$ and zero potential energy. Show that equations of motion for this system are given by Euler's equations:

$$\dot{\Omega} = ad^* v.\Omega$$

where $v = g^{-1}\dot{g} \in \mathfrak{g}$, $\Omega = Av \in \mathfrak{g}^*$, and ad^* is the coadjoint action:

$$\langle ad^* x.f, y \rangle = -\langle f, ad x.y \rangle \qquad x, y \in \mathfrak{g}, f \in \mathfrak{g}^*.$$

(For $G = SO(3, \mathbb{R})$, this system describes motion of a solid body rotating around its center of gravity – so called Euler's case of rotation of a solid body. In this case, A describes the body's moment of inertia, v is angular velocity, and Ω is angular momentum, both measured in the moving frame. Details can be found in [1]).

(2) Using the results of the previous part, show that if A is a bi-invariant metric on G, then one-parameter subgroups $\exp(tx), x \in \mathfrak{g}$ are geodesics for this metric.

3.15. Let G be a complex connected simply-connected Lie group, with Lie algebra $\mathfrak{g} = \mathrm{Lie}(G)$, and let $\mathfrak{k} \subset \mathfrak{g}$ be a real form of $\mathfrak{g}$.

(1) Define the $\mathbb{R}$-linear map $\theta \colon \mathfrak{g} \to \mathfrak{g}$ by $\theta(x + iy) = x - iy$, $x, y \in \mathfrak{k}$. Show that θ is an automorphism of $\mathfrak{g}$ (considered as a real Lie algebra), and that it can be uniquely lifted to an automorphism $\theta \colon G \to G$ of the group G (considered as a real Lie group).

(2) Let $K = G^\theta$. Show that K is a real Lie group with Lie algebra $\mathfrak{k}$.

3.16. Let $\mathrm{Sp}(n)$ be the unitary quaternionic group defined in Section 2.7. Show that $\mathfrak{sp}(n)_{\mathbb{C}} = \mathfrak{sp}(n, \mathbb{C})$. Thus $\mathrm{Sp}(n)$ is a compact real form of $\mathrm{Sp}(n, \mathbb{C})$.

3.17. Let $\mathfrak{so}(p,q) = \mathrm{Lie}(\mathrm{SO}(p,q))$. Show that its complexification is $\mathfrak{so}(p,q)_{\mathbb{C}} = \mathfrak{so}(p+q, \mathbb{C})$.

3.18. Let

$$S = \begin{pmatrix} 0 & -1 \\ 1 & 0 \end{pmatrix} \in \mathrm{SL}(2, \mathbb{C}).$$

(1) Show that $S = \exp\left(\frac{\pi}{2}(f - e)\right)$, where $e, f \in \mathfrak{sl}(2, \mathbb{C})$ are defined by (3.23).

(2) Compute $\mathrm{Ad}\, S$ in the basis e, f, h.

3.19. Let G be a complex connected Lie group.

(1) Show that $g \mapsto \mathrm{Ad}\, g$ is an analytic map $G \to \mathfrak{gl}(\mathfrak{g})$.

(2) Assume that G is compact. Show that then $\mathrm{Ad}\, g = 1$ for any $g \in G$.

(3) Show that any connected compact complex group must be commutative.

(4) Show that if G is a connected complex compact group, then the exponential map gives an isomorphism of Lie groups

$$\mathfrak{g}/L \simeq G$$

for some lattice $L \subset \mathfrak{g}$ (i.e. a free abelian group of rank equal to $2 \dim \mathfrak{g}$).

4

Representations of Lie groups and Lie algebras

In this section, we will discuss the representation theory of Lie groups and Lie algebras – as far as it can be discussed without using the structure theory of semisimple Lie algebras. Unless specified otherwise, all Lie groups, algebras, and representations are finite-dimensional, and all representations are complex. Lie groups and Lie algebras can be either real or complex; unless specified otherwise, all results are valid both for the real and complex case.

4.1. Basic definitions

Let us start by recalling the basic definitions.

Definition 4.1. A representation of a Lie group G is a vector space V together with a morphism $\rho: G \to \mathrm{GL}(V)$.

A representation of a Lie algebra $\mathfrak{g}$ is a vector space V together with a morphism $\rho: \mathfrak{g} \to \mathfrak{gl}(V)$.

A morphism between two representations V, W of the same group G is a linear map $f: V \to W$ which commutes with the action of $G: f\rho(g) = \rho(g)f$. In a similar way one defines a morphism of representations of a Lie algebra. The space of all G-morphisms (respectively, $\mathfrak{g}$-morphisms) between V and W will be denoted by $\mathrm{Hom}_G(V, W)$ (respectively, $\mathrm{Hom}_{\mathfrak{g}}(V, W)$).

Remark 4.2. Morphisms between representations are also frequently called *intertwining operators* because they "intertwine" action of G in V and W.

The notion of a representation is completely parallel to the notion of module over an associative ring or algebra; the difference of terminology is due to historical reasons. In fact, it is also usual to use the word "module" rather than "representation" for Lie algebras: a module over Lie algebra $\mathfrak{g}$ is the same as a representation of $\mathfrak{g}$. We will use both terms interchangeably.

Note that in this definition we did not specify whether V and G, $\mathfrak{g}$ are real or complex. Usually if G (respectively, $\mathfrak{g}$) is complex, then V should also be taken a complex vector space. However, it also makes sense to take complex V even if G is real: in this case we require that the morphism $G \to \mathrm{GL}(V)$ be smooth, considering $\mathrm{GL}(V)$ as $2n^2$-dimensional real manifold. Similarly, for real Lie algebras we can consider complex representations requiring that $\rho\colon \mathfrak{g} \to \mathfrak{gl}(V)$ be $\mathbb{R}$-linear.

Of course, we could also restrict ourselves to consideration of real representations of real groups. However, it turns out that the introduction of complex representations significantly simplifies the theory even for real groups and algebras. Thus, from now on, all representations will be complex unless specified otherwise.

The first important result about representations of Lie groups and Lie algebras is the following theorem.

Theorem 4.3. *Let G be a Lie group (real or complex) with Lie algebra $\mathfrak{g}$.*

(1) *Every representation $\rho\colon G \to \mathrm{GL}(V)$ defines a representation $\rho_*\colon \mathfrak{g} \to \mathfrak{gl}(V)$, and every morphism of representations of G is automatically a morphism of representations of $\mathfrak{g}$.*

(2) *If G is connected, simply-connected, then $\rho \mapsto \rho_*$ gives an equivalence of categories of representations of G and representations of $\mathfrak{g}$. In particular, every representation of $\mathfrak{g}$ can be uniquely lifted to a representation of G, and $\mathrm{Hom}_G(V, W) = \mathrm{Hom}_\mathfrak{g}(V, W)$.*

Indeed, part (1) is a special case of Theorem 3.20, and part (2) follows from Theorem 3.41.

This is an important result, as Lie algebras are, after all, finite dimensional vector spaces, so they are easier to deal with. For example, this theorem shows that a representation of $\mathrm{SU}(2)$ is the same as a representation of $\mathfrak{su}(2)$, i.e. a vector space with three endomorphisms X, Y, Z, satisfying commutation relations $XY - YX = Z, YZ - ZY = X, ZX - XZ = Y$.

This theorem can also be used to describe representations of a group which is connected but not simply-connected: indeed, by Corollary 3.43 any such group can be written as $G = \tilde{G}/Z$ for some simply-connected group $\tilde{G}$ and a discrete central subgroup $Z \subset G$. Thus, representations of G are the same as representations of $\tilde{G}$ satisfying $\rho(Z) = \mathrm{id}$. An important example of this is when $G = \mathrm{SO}(3, \mathbb{R})$, $\tilde{G} = \mathrm{SU}(2)$ (see Exercise 4.1).

Lemma 4.4. *Let $\mathfrak{g}$ be a real Lie algebra, and $\mathfrak{g}_\mathbb{C}$ its complexification as defined in Definition 3.49. Then any complex representation of $\mathfrak{g}$ has a unique structure* *of representation of* $\mathfrak{g}_\mathbb{C}$, *and* $\mathrm{Hom}_\mathfrak{g}(V, W) = \mathrm{Hom}_{\mathfrak{g}_\mathbb{C}}(V, W)$. *In other words, categories of complex representations of* $\mathfrak{g}$, $\mathfrak{g}_\mathbb{C}$ *are equivalent.*

Proof. Let $\rho\colon \mathfrak{g} \to \mathfrak{gl}(V)$ be the representation of $\mathfrak{g}$. Extend it to $\mathfrak{g}_\mathbb{C}$ by $\rho(x + iy) = \rho(x) + i\rho(y)$. We leave it to the reader to check that so defined ρ is complex-linear and agrees with the commutator. $\square$

Example 4.5. The categories of finite-dimensional representations of $\mathrm{SL}(2, \mathbb{C})$, $\mathrm{SU}(2)$, $\mathfrak{sl}(2, \mathbb{C})$ and $\mathfrak{su}(2)$ are all equivalent. Indeed, by results of Section 3.10, $\mathfrak{sl}(2, \mathbb{C}) = (\mathfrak{su}(2))_\mathbb{C}$, so categories of their finite-dimensional representations are equivalent; since Lie groups $\mathrm{SU}(2)$, $\mathrm{SL}(2, \mathbb{C})$ are simply-connected, they have the same representations as the corresponding Lie algebras.

This, in particular, allows us to reduce the problem of study of representations of a non-compact Lie group $\mathrm{SL}(2, \mathbb{C})$ to the study of representations of a compact Lie group $\mathrm{SU}(2)$. This is useful because, as we will show below, representation theory of compact Lie groups is especially nice.

Remark 4.6. This only works for finite-dimensional representations; the theory of infinite-dimensional representations of $\mathrm{SL}(2, \mathbb{C})$ is very different from that of $\mathrm{SU}(2)$.

The following are some examples of representations that can be defined for any Lie group G (and thus, for any Lie algebra $\mathfrak{g}$).

Example 4.7. Trivial representation: $V = \mathbb{C}$, $\rho(g) = \mathrm{id}$ for any $g \in G$ (respectively, $\rho(x) = 0$ for any $x \in \mathfrak{g}$).

Example 4.8. Adjoint representation: $V = \mathfrak{g}$, $\rho(g) = \mathrm{Ad}\, g$ (respectively, $\rho(x) = \mathrm{ad}\, x$). See (2.4), Lemma 3.15 for definition of Ad, ad.

4.2. Operations on representations

In this section, we discuss basic notions of representation theory of Lie groups and Lie algebras, giving examples of representations, operations on representations such as direct sum and tensor product, and more.

4.2.1. Subrepresentations and quotients

Definition 4.9. Let V be a representation of G (respectively $\mathfrak{g}$). A subrepresentation is a vector subspace $W \subset V$ stable under the action: $\rho(g)W \subset W$ for all $g \in G$ (respectively, $\rho(x)W \subset W$ for all $x \in \mathfrak{g}$).

It is easy to check that if G is a connected Lie group with Lie algebra $\mathfrak{g}$, then $W \subset V$ is a subrepresentation for G if and only if it is a subrepresentation for $\mathfrak{g}$.

It is trivial to check that if $W \subset V$ is a subrepresentation, then the quotient space V/W has a canonical sructure of a representation. It will be called *factor representation*, or the *quotient representation*.

4.2.2. Direct sum and tensor product

Lemma 4.10. *Let V, W be representations of G (respectively, $\mathfrak{g}$). Then there is a canonical structure of a representation on $V^*, V \oplus W, V \otimes W$.*

Proof. Action of G on $V \oplus W$ is given by $\rho(g)(v + w) = \rho(g)v + \rho(g)w$ (for $g \in G$, $v \in V$, $w \in W$) and similarly for $\mathfrak{g}$.

For tensor product, we define $\rho(g)(v \otimes w) = \rho(g)v \otimes \rho(g)w$. However, the action of $\mathfrak{g}$ is trickier: indeed, naive definition $\rho(x)(v \otimes w) = \rho(x)v \otimes \rho(x)w$ (for $x \in \mathfrak{g}$) does not define a representation (it is not even linear in x). Instead, if we write $x = \dot{\gamma}(0)$ for some one-parameter subgroup $\gamma(t)$ in the Lie group G with $\gamma(0) = 1$, then

$$\rho(x)(v \otimes w) = \frac{\mathrm{d}}{\mathrm{d}t}|_{t=0}(\gamma(t)v \otimes \gamma(t)w) = (\dot{\gamma}(0)v \otimes \gamma(0)w)$$
$$+ (\gamma(0)v \otimes \dot{\gamma}(t)w)$$
$$= \rho(x)v \otimes w + v \otimes \rho(x)w$$

by using Leibniz rule. Thus, we define for $x \in \mathfrak{g}$

$$\rho(x)(v \otimes w) = \rho(x)v \otimes w + v \otimes \rho(x)w.$$

It is easy to show, even without using the Lie group G, that so defined action is indeed a representation of $\mathfrak{g}$ on $V \otimes W$.

To define the action of G, $\mathfrak{g}$ on V^*, we require that the natural pairing $V \otimes V^* \to \mathbb{C}$ be a morphism of representations, considering $\mathbb{C}$ as the trivial representation. This gives, for $v \in V, v^* \in V^*$, $\langle \rho(g)v, \rho(g)v^* \rangle = \langle v, v^* \rangle$, so the action of G in V^* is given by $\rho_{V^*}(g) = \rho(g^{-1})^t$, where for $A \colon V \to V$, we denote by A^t the adjoint operator $V^* \to V^*$.

Similarly, for the action of $\mathfrak{g}$ we get $\langle \rho(x)v, v^* \rangle + \langle v, \rho(x)v^* \rangle = 0$, so $\rho_{V^*}(x) = -(\rho_V(x))^t$. $\qquad\square$

As an immediate corollary, we see that for a representation V, any tensor space $V^{\otimes k} \otimes (V^*)^{\otimes l}$ has a canonical structure of a representation.

Example 4.11. For any Lie algebra $\mathfrak{g}$, the dual vector space $\mathfrak{g}^*$ has a canonical structure of a representation of $\mathfrak{g}$, given by

$$\langle \mathrm{ad}^* x.f, y \rangle = -\langle f, \mathrm{ad}\, x.y \rangle, \qquad f \in \mathfrak{g}^*, x, y \in \mathfrak{g}.$$

This representation is called the *coadjoint representation* and plays an important role in representation theory: for example, for a large class of Lie algebras there is a bijection between (some) G–orbits in $\mathfrak{g}^*$ and finite-dimensional irreducible representations of $\mathfrak{g}$ (see [30]).

Example 4.12. Let V be a representation of G (respectively, $\mathfrak{g}$). Then the space $\mathrm{End}(V) \simeq V \otimes V^*$ of linear operators on V is also a representation, with the action given by $g \colon A \mapsto \rho_V(g) A \rho_V(g^{-1})$ for $g \in G$ (respectively, $x \colon A \mapsto \rho_V(x) A - A \rho_V(x)$ for $x \in \mathfrak{g}$). More generally, the space of linear maps $\mathrm{Hom}(V, W)$ between two representations is also a representation with the action defined by $g \colon A \mapsto \rho_W(g) A \rho_V(g^{-1})$ for $g \in G$ (respectively, $x \colon A \mapsto \rho_W(x) A - A \rho_V(x)$ for $x \in \mathfrak{g}$).

Similarly, the space of bilinear forms on V is also a representation, with action given by

$$gB(v, w) = B(g^{-1}v, g^{-1}w), \quad g \in G$$
$$xB(v, w) = -(B(x.v, w) + B(v, x.w)), \quad x \in \mathfrak{g}.$$

Proof of these formulas is left to the reader as an exercise.

4.2.3. Invariants

Definition 4.13. Let V be a representation of a Lie group G. A vector $v \in V$ is called *invariant* if $\rho(g)v = v$ for all $g \in G$. The subspace of invariant vectors in V is denoted by V^G.

Similarly, let V be a representation of a Lie algebra $\mathfrak{g}$. A vector $v \in V$ is called *invariant* if $\rho(x)v = 0$ for all $x \in \mathfrak{g}$. The subspace of invariant vectors in V is denoted by $V^{\mathfrak{g}}$.

We leave it to the reader to prove that if G is a connected Lie group with the Lie algebra $\mathfrak{g}$, then for any representation V of G, we have $V^G = V^{\mathfrak{g}}$.

Example 4.14. Let V, W be representations and $\mathrm{Hom}(V, W)$ be the space of linear maps $V \to W$, with the action of G defined as in Example 4.12. Then $(\mathrm{Hom}(V, W))^G = \mathrm{Hom}_G(V, W)$ is the space of intertwining operators. In particular, this shows that $V^G = (\mathrm{Hom}(\mathbb{C}, V))^G = \mathrm{Hom}_G(\mathbb{C}, V)$, with $\mathbb{C}$ considered as a trivial representation.

Example 4.15. Let B be a bilinear form on a representation V. Then B is invariant under the action of G defined in Example 4.12 iff

$$B(gv, gw) = B(v, w)$$

for any $g \in G$, $v, w \in V$. Similarly, B is invariant under the action of $\mathfrak{g}$ iff

$$B(x.v, w) + B(v, x.w) = 0$$

for any $x \in \mathfrak{g}$, $v, w \in V$.

We leave it to the reader to check that B is invariant iff the linear map $V \to V^*$ defined by $v \mapsto B(v, -)$ is a morphism of representations.

4.3. Irreducible representations

One of the main problems of the representation theory is the problem of classification of all representations of a Lie group or a Lie algebra. In this generality, it is an extremely difficult problem and for a general Lie group, no satisfactory answer is known. We will later show that for some special classes of Lie groups (namely compact Lie groups and semisimple Lie groups, to be defined later) this problem does have a good answer.

The most natural approach to this problem is to start by studying the simplest possible representations, which would serve as building blocks for more complicated representations.

Definition 4.16. A non-zero representation V of G or $\mathfrak{g}$ is called *irreducible* or *simple* if it has no subrepresentations other than 0, V. Otherwise V is called *reducible*.

Example 4.17. Space $\mathbb{C}^n$, considered as a representation of $SL(n, \mathbb{C})$, is irreducible.

If a representation V is reducible, then it has a non-trivial subrepresentation W and thus, V can be included in a short exact sequence $0 \to W \to V \to V/W \to 0$; thus, in a certain sense it is built out of simpler pieces. The natural question is whether this exact sequence splits, i.e. whether we can write $V = W \oplus (V/W)$ as a representation. If so then repeating this process, we can write V as a direct sum of irreducible representations.

Definition 4.18. A representation is called *completely reducible* or *semisimple* if it is isomorphic to a direct sum of irreducible representations: $V \simeq \bigoplus V_i$, V_i irreducible.

In such a case one usually groups together isomorphic summands writing $V \simeq \bigoplus n_i V_i$, $n_i \in \mathbb{Z}_+$, where V_i are pairwise non-isomorphic irreducible representations. The numbers n_i are called *multiplicities*.

However, as the following example shows, not every representation is completely reducible.

Example 4.19. Let $G = \mathbb{R}$, so $\mathfrak{g} = \mathbb{R}$. Then a representation of $\mathfrak{g}$ is the same as a vector space V with a linear map $\mathbb{R} \to \text{End}(V)$; obviously, every such map is of the form $t \mapsto tA$ for some $A \in \text{End}(V)$ which can be arbitrary. The corresponding representation of the group $\mathbb{R}$ is given by $t \mapsto \exp(tA)$. Thus, classifying representations of $\mathbb{R}$ is equivalent to classifying linear maps $V \to V$ up to a change of basis. Such a classification is known (Jordan normal form) but non-trivial.

If v is an eigenvector of A then the one-dimensional space $\mathbb{C}v \subset V$ is invariant under A and thus a subrepresentation in V. Since every linear operator in a complex vector space has an eigenvector, this shows that every representation of $\mathbb{R}$ is reducible, unless it is one-dimensional. Thus, the only irreducible representations of $\mathbb{R}$ are one-dimensional.

Now one easily sees that writing a representation given by $t \mapsto \exp(tA)$ as a direct sum of irreducible ones is equivalent to diagonalizing A. So a representation is completely reducible iff A is diagonalizable. Since not every linear operator is diagonalizable, not every representation is completely reducible.

Thus, more modest goals of the representation theory would be answering the following questions:

(1) For a given Lie group G, classify all irreducible representations of G.
(2) For a given representation V of a Lie group G, given that it is completely reducible, find explicitly the decomposition of V into direct sum of irreducibles.
(3) For which Lie groups G all representations are completely reducible?

One tool which can be used in decomposing representations into direct sum is the use of central elements.

Lemma 4.20. *Let* $\rho: G \to \text{GL}(V)$ *be a representation of* G *(respectively,* $\mathfrak{g}$*) and* $A: V \to V$ *a diagonalizable intertwining operator. Let* $V_\lambda \subset V$ *be the eigenspace for* A *with eigenvalue* λ*. Then each* V_λ *is a subrepresentation, so* $V = \bigoplus V_\lambda$ *as a representation of* G *(respectively* $\mathfrak{g}$*).*

The proof of this lemma is trivial and is left to the reader. As an immediate corollary, we get the following result.

Lemma 4.21. *Let* V *be a representation of* G *and let* $Z \in Z(G)$ *be a central element of* G *such that* $\rho(Z)$ *is diagonalizable. Then as a representation of* G*,* $V = \bigoplus V_\lambda$*, where* V_λ *is the eigenspace for* $\rho(Z)$ *with eigenvalue* λ*. Similar result also holds for central elements in* $\mathfrak{g}$*.*

Of course, there is no guarantee that V_λ will be an irreducible representation; moreover, in many cases the Lie groups we consider have no central elements at all.

Example 4.22. Consider action of $GL(n, \mathbb{C})$ on $\mathbb{C}^n \otimes \mathbb{C}^n$. Then the permutation operator $P: v \otimes w \mapsto w \otimes v$ commutes with the action of $GL(n, \mathbb{C})$, so the subspaces $S^2\mathbb{C}^n$, $\Lambda^2\mathbb{C}^n$ of symmetric and skew-symmetric tensors (which are exactly the eigenspaces of P) are $GL(n, \mathbb{C})$-invariant and $\mathbb{C}^n \otimes \mathbb{C}^n = S^2\mathbb{C}^n \oplus \Lambda^2\mathbb{C}^n$ as a representation. In fact, both $S^2\mathbb{C}^n$, $\Lambda^2\mathbb{C}^n$ are irreducible (this is not obvious but can be proved by a lengthy explicit calculation; a better way of proving this will be given in Exercise 8.4). Thus, $\mathbb{C}^n \otimes \mathbb{C}^n$ is completely reducible.

4.4. Intertwining operators and Schur's lemma

Recall that an intertwining operator is a linear map $V \to W$ which commutes with the action of G. Such operators frequently appear in various applications. A typical example is quantum mechanics, where we have a vector space V (describing the state space of some mechanical system) and the Hamiltonian operator $H: V \to V$. Then saying that this system has a symmetry described by a group G is the same as saying that we have an action of G on V which leaves H invariant, i.e. $gHg^{-1} = H$ for any $g \in G$. This exactly means that H is an intertwining operator. A real-life example of such situation (spherical Laplace operator) will be described in Section 4.9.

These examples motivate the study of intertwining operators. For example, does G-invariance of an operator helps computing eigenvalues and eigenspaces?

The first result in this direction is the following famous lemma.

Lemma 4.23. (Schur Lemma)

(1) *Let V be an irreducible complex representation of G. Then the space of intertwining operators $\mathrm{Hom}_G(V, V) = \mathbb{C} \,\mathrm{id}$: any endomorphism of an irreducible representation of G is constant.*
(2) *If V and W are irreducible complex representations which are not isomorphic then $\mathrm{Hom}_G(V, W) = 0$.*

Similar result holds for representations of a Lie algebra $\mathfrak{g}$.

Proof. We note that if $\Phi: V \to W$ is an intertwining operator, then $\mathrm{Ker}\,\Phi$, $\mathrm{Im}\,\Phi$ are subrepresentations in V, W, respectively. If V is irreducible, either $\mathrm{Ker}\,\Phi = V$ (in which case $\Phi = 0$) or $\mathrm{Ker}\,\Phi = 0$, so Φ is injective.

Similarly, if W is irreducible, either Im $\Phi = 0$ (so $\Phi = 0$) or Im $\Phi = W$, Φ is surjective. Thus, either $\Phi = 0$ or Φ is an isomorphism.

Now part (2) follows immediately: since V, W are not isomorphic, Φ must be zero. To prove part (1), notice that the above argument shows that any non-zero intertwiner $V \to V$ is invertible. Now let λ be an eigenvalue of Φ. Then $\Phi - \lambda \operatorname{id}$ is not invertible. On the other hand, it is also an intertwiner, so it must be zero. Thus, $\Phi = \lambda \operatorname{id}$. $\square$

Example 4.24. Consider the group $GL(n, \mathbb{C})$. Since $\mathbb{C}^n$ is irreducible as a representation of $GL(n, \mathbb{C})$, every operator commuting with $GL(n, \mathbb{C})$ must be scalar. Thus, the center $Z(GL(n, \mathbb{C})) = \{\lambda \cdot \operatorname{id}, \lambda \in \mathbb{C}^\times\}$; similarly, the center of the Lie algebra is $\mathfrak{z}(\mathfrak{gl}(n, \mathbb{C})) = \{\lambda \cdot \operatorname{id}, \lambda \in \mathbb{C}\}$.

Since $\mathbb{C}^n$ is also irreducible as a representation of $SL(n, \mathbb{C})$, $U(n)$, $SU(n)$, $SO(n, \mathbb{C})$, a similar argument can be used to compute the center of each of these groups. The answer is

$$Z(SL(n, \mathbb{C})) = Z(SU(n)) = \{\lambda \cdot \operatorname{id}, \quad \lambda^n = 1\} \quad \mathfrak{z}(\mathfrak{sl}(n, \mathbb{C})) = \mathfrak{z}(\mathfrak{su}(n)) = 0$$

$$Z(U(n)) = \{\lambda \cdot \operatorname{id}, \quad |\lambda| = 1\} \qquad\qquad \mathfrak{z}(\mathfrak{u}(n)) = \{\lambda \cdot \operatorname{id}, \quad \lambda \in i\mathbb{R}\}$$

$$Z(SO(n, \mathbb{C})) = Z(SO(n, \mathbb{R})) = \begin{cases} \{\pm 1\} & n \text{ even}, n > 2 \\ \{1\} & n \text{ odd} \end{cases} \quad \mathfrak{z}(\mathfrak{so}(n, \mathbb{C})) = \mathfrak{z}(\mathfrak{so}(n, \mathbb{R})) = 0.$$

As an immediate corollary of Schur's lemma, we get the following result.

Corollary 4.25. *Let V be a completely reducible representation of Lie group G (respectively, Lie algebra $\mathfrak{g}$). Then*

(1) *If $V = \bigoplus V_i$, V_i – irreducible, pairwise non-isomorphic, then any inter-twining operator $\Phi : V \to V$ is of the form $\Phi = \bigoplus \lambda_i \operatorname{id}_{V_i}$.*

(2) *If $V = \bigoplus n_i V_i = \bigoplus \mathbb{C}^{n_i} \otimes V_i$, V_i – irreducible, pairwise non-isomorphic, then any intertwining operator $\Phi : V \to V$ is of the form $\Phi = \bigoplus (A_i \otimes \operatorname{id}_{V_i})$, $A_i \in \operatorname{End}(\mathbb{C}^{n_i})$.*

Proof. For part (1), notice that any operator $V \to V$ can be written in a block form: $\Phi = \bigoplus \Phi_{ij}$, $\Phi_{ij} : V_i \to V_j$. By Schur's lemma, $\Phi_{ij} = 0$ for $i \neq j$ and $\Phi_{ii} = \lambda_i \operatorname{id}_{V_i}$. Part (2) is proved similarly. $\square$

This result shows that indeed, if we can decompose a representation V into irreducible ones, this will give us a very effective tool for analysing intertwining operators. For example, if $V = \bigoplus V_i, V_i \not\simeq V_j$, then $\Phi = \bigoplus \lambda_i \operatorname{id}_{V_i}$, so one can find λ_i by computing $\Phi(v)$ for just one vector in V_i. It also shows that each eigenvalue λ_i will appear with multiplicity equal to $\dim V_i$. This is exactly what we did in the baby example in the introduction, where we had $G = \mathbb{Z}_n$.

Another useful corollary of Schur's lemma is the following result.

Proposition 4.26. *If G is a commutative group, then any irreducible complex representation of G is one-dimensional. Similarly, if $\mathfrak{g}$ is a commutative Lie algebra, then any irreducible complex representation of $\mathfrak{g}$ is one-dimensional.*

Indeed, since G is commutative, every $\rho(g)$ commutes with the action of G, so $\rho(g) = \lambda(g)\,\mathrm{id}$.

Example 4.27. Let $G = \mathbb{R}$. Then its irreducible representations are one-dimensional (this had already been dicussed before, see Example 4.19). In fact, it is easy to describe them: one-dimensional representations of the corresponding Lie algebra $\mathfrak{g} = \mathbb{R}$ are $a \mapsto \lambda a$, $\lambda \in \mathbb{C}$. Thus, irreducible representations of $\mathbb{R}$ are V_λ, $\lambda \in \mathbb{C}$, where each V_λ is a one-dimensional complex space with the action of $\mathbb{R}$ given by $\rho(a) = e^{\lambda a}$.

In a similar way, one can describe irreducible representations of $S^1 = \mathbb{R}/\mathbb{Z}$: they are exactly those representations of $\mathbb{R}$ which satisfy $\rho(a) = 1$ for $a \in \mathbb{Z}$. Thus, irreducible representations of S^1 are V_k, $k \in \mathbb{Z}$, where each V_k is a one-dimensional complex space with the action of S^1 given by $\rho(a) = e^{2\pi i k a}$. In the realization $S^1 = \{z \in \mathbb{C} \mid |z| = 1\}$ the formula is even simpler: in V_k, z acts by z^k.

4.5. Complete reducibility of unitary representations: representations of finite groups

In this section, we will show that a large class of representations is completely reducible.

Definition 4.28. A complex representation V of a real Lie group G is called unitary if there is a G-invariant inner product: $(\rho(g)v, \rho(g)w) = (v, w)$, or equivalently, $\rho(g) \in U(V)$ for any $g \in G$. (The word "inner product" means a positive definite Hermitian form.)

Similarly, a representation V of a real Lie algebra $\mathfrak{g}$ is called unitary if there is an inner product which is $\mathfrak{g}$-invariant: $(\rho(x)v, w) + (v, \rho(x)w) = 0$, or equivalently, $\rho(x) \in \mathfrak{u}(V)$ for any $x \in \mathfrak{g}$.

Example 4.29. Let $V = F(S)$ be the space of complex-valued functions on a finite set S. Let G be a finite group acting by permutations on S; then it also acts on V by (2.1). Then $(f_1, f_2) = \sum_s f_1(s)\overline{f_2(s)}$ is an invariant inner product, so such a representation is unitary.

The following result explains why unitary representations are so important.

Theorem 4.30. *Each unitary representation is completely reducible.*

Proof. The proof goes by induction on dimension. Either V is irreducible, and we're done, or V has a subrepresentation W. Then $V = W \oplus W^{\perp}$, and $W^{\perp}$ is a subrepresentation as well. Indeed: if $w \in W^{\perp}$, then $(gw, v) = (w, g^{-1}v) = 0$ for any $v \in W$ (since $g^{-1}v \in W$), so $gw \in W^{\perp}$. A similar argument applies to representations of Lie algebras. $\qquad\square$

Theorem 4.31. *Any representation of a finite group is unitary.*

Proof. Let $B(v, w)$ be some inner product in V. Of course, it may not be G-invariant, so $B(gv, gw)$ may be different from $B(v, w)$. Let us "average" B by using group action:

$$\tilde{B}(v, w) = \frac{1}{|G|} \sum_{g \in G} B(gv, gw).$$

Then $\tilde{B}$ is positive definite (it is a sum of positive definite forms), and it is G-invariant:

$$\tilde{B}(hv, hw) = \frac{1}{|G|} \sum_{g \in G} B(ghv, ghw) = \frac{1}{|G|} \sum_{g' \in G} B(g'v, g'w)$$

by making subsitution $gh = g'$ and noticing that as g runs over G, so does g'. $\quad\square$

Combining this with Theorem 4.30, we immediately get the main result of this section.

Theorem 4.32. *Every representation of a finite group is completely reducible.*

Note that this theorem does not give an explicit recipe for decomposing a representation into direct sum of irreducibles. We will return to this problem later.

4.6. Haar measure on compact Lie groups

In the previous section we have proved complete reducibility of representations of a finite group G. The natural question is whether this proof can be generalized to Lie groups.

Analyzing the proof, we see that the key step was averaging a function over the group: $\tilde{f} = (1/|G|) \sum f(g)$ for a complex-valued function on a group. It seems reasonable to expect that in the case of Lie groups, we should replace the sum by suitably defined integral over G.

Definition 4.33. A right Haar measure on a real Lie group G is a Borel measure dg which is invariant under the right action of G on itself.

Right invariance implies (and, in fact, is equivalent to) the identity $\int f(gh)\,dg = \int f(g)\,dg$ for any $h \in G$ and integrable function f. In a similar way one defines left Haar measure on G.

To construct such a measure, we start by constructing an invariant volume form on G.

Theorem 4.34. *Let G be a real Lie group.*

(1) *G is orientable; moreover, orientation can be chosen so that the right action of G on itself preserves the orientation.*
(2) *If G is compact, then for a fixed choice of right-invariant orientation on G there exists a unique right-invariant top degree differential form ω such that $\int_G \omega = 1$.*
(3) *The form ω is left-invariant only if G is connected. Otherwise, ω is left invariant up to a sign.*

Proof. Let us choose some non-zero element in $\Lambda^n \mathfrak{g}^*$, $n = \dim G$. Then it can be uniquely extended to a right-invariant differential form $\tilde{\omega}$ on G (see Theorem 2.27). Since this form is non-vanishing on G, this shows that G is orientable.

If G is compact, the integral $I = \int_G \tilde{\omega}$ is finite. Define $\omega = \tilde{\omega}/I$. Then ω is right-invariant and satisfies $\int_G \omega = 1$, thus proving existence statement of part (2). Uniqueness is obvious: by Theorem 2.27, space of right-invariant forms is identified with $\Lambda^n \mathfrak{g}^*$, which is one-dimensional; thus any right-invariant form ω' is has the form $c\omega$, and $\int_G \omega' = 1$ implies $c = 1$.

To prove that ω is also left-invariant, it suffices to check that it is invariant under coadjoint action (cf. Theorem 2.28). But $\Lambda^n \mathfrak{g}^*$ is a one-dimensional representation of G. Thus, this result immediately follows from the following lemma.

Lemma 4.35. *Let V be a one-dimensional real representation of a compact Lie group G. Then for any $g \in G$, $|\rho(g)| = 1$.*

Indeed, if $|\rho(g)| < 1$, then $\rho(g^n) \to 0$ as $n \to \infty$. But $\rho(G)$ is a compact subset in $\mathbb{R}^\times$, so it can not contain a sequence with limit 0. In a similar way, $|\rho(g)| > 1$ also leads to a contradiction.

To prove invariance under $i\colon g \mapsto g^{-1}$, notice that since ω is left-invariant, it is easy to see that $i^*(\omega)$ is a right-invariant form on G; thus, it suffices to check that ω and $i^*(\omega)$ are equal up to a sign at $1 \in G$. But $i_*\colon \mathfrak{g} \to \mathfrak{g}$ is given by $x \mapsto -x$ (which follows from $i(\exp(tx)) = \exp(-tx)$). Thus, on $\Lambda^n \mathfrak{g}$, $i_* = (-1)^n$, so $i^*(\omega) = (-1)^n \omega$. $\quad\square$

We can now prove existence of bi-invariant measure on compact Lie groups.

Theorem 4.36. *Let G be a compact real Lie group. Then it has a canonical Borel measure* dg *which is both left- and right-invariant and invariant under* $g \mapsto g^{-1}$ *and which satisfies* $\int_G dg = 1$. *This measure is called the Haar measure on G and is usually denoted by* dg.

Proof. Choose an orientation of G and a bi-invariant volume form ω as in Theorem 4.34. Then general results of measure theory imply that there exists a unique Borel measure dg on G such that for any continuous function f, we have $\int_G f \, dg = \int_G f \, \omega$. Invariance of dg under left and right action and under $g \mapsto g^{-1}$ follows from invariance of ω. $\quad\square$

It is not difficult to show that the Haar measure is unique (see, e.g., [32, Section VIII.2]).

Remark 4.37. In fact, bi-invariant Haar measure exists not only for every Lie group but also for every compact topological group (with some technical restrictions). However, in full generality this result is much harder to prove.

Example 4.38. Let $G = S^1 = \mathbb{R}/\mathbb{Z}$. Then the Haar measure is the usual measure dx on $\mathbb{R}/\mathbb{Z}$.

Note that in general, explicitly writing the Haar measure on a group is not easy – for example, because in general there is no good choice of coordinates on G. Even in those cases when a coordinate system on G can be described explicitly, the Haar measure is usually given by rather complicated formulas. The only case where this measure can be written by a formula simple enough to be useful for practical computations is when we integrate conjugation-invariant functions (also called class functions).

Example 4.39. Let $G = U(n)$ and let f be a smooth function on G such that $f(ghg^{-1}) = f(h)$. Then

$$\int_{U(n)} f(g) dg = \frac{1}{n!} \int_T f \begin{pmatrix} t_1 & & \\ & t_2 & \\ & & \ddots \\ & & & t_n \end{pmatrix} \prod_{i<j} |t_i - t_j|^2 dt,$$

where

$$T = \left\{ \begin{pmatrix} t_1 & & & \\ & t_2 & & \\ & & \ddots & \\ & & & t_n \end{pmatrix}, t_k = e^{i\varphi_k} \right\}$$

is the subgroup of diagonal matrices and $dt = \frac{1}{(2\pi)^n} d\varphi_1 \dots d\varphi_n$ is the Haar measure on T.

This is a special case of *Weyl Integration Formula*. The statement of this theorem in full generality and the proof can be found, for example, in [5] or in [32]. The proof requires a fair amount of the structure theory of compact Lie groups and will not be given here.

The main result of this section is the following theorem.

Theorem 4.40. *Any finite-dimensional representation of a compact Lie group is unitary and thus completely reducible.*

Proof. The proof is almost identical to the proof for the finite group: let $B(v, w)$ be some positive definite inner product in V and "average" it by using group action:

$$\tilde{B}(v, w) = \int_G B(gv, gw) \, dg,$$

where dg is the Haar measure on G. Then $\tilde{B}(v, v) > 0$ (it is an integral of a positive function) and right invariance of Haar measure shows that $\tilde{B}(hv, hw) = \tilde{B}(v, w)$.

Complete reducibility now follows from Theorem 4.30. $\qquad\square$

4.7. Orthogonality of characters and Peter–Weyl theorem

In the previous section, we have established that any representation of a compact Lie group is completely reducible: $V \simeq \bigoplus n_i V_i$, where $n_i \in \mathbb{Z}_+, V_i$ are pairwise non-isomorphic irreducible representations. However, we have not yet discussed how one can explicitly decompose a given representation in a direct sum of irreducibles, or at least find the multiplicities n_i. This will be discussed in this section. Throughout this section, G is a compact real Lie group with Haar measure dg.

Let v_i be a basis in a representation V. Writing the operator $\rho(g) \colon V \to V$ in the basis v_i, we get a matrix-valued function on G. Equivalently, we can consider each entry $\rho_{ij}(g)$ as a scalar-valued function on G. Such functions are called *matrix coefficients* (of the representation V).

Theorem 4.41.

(1) *Let V, W be non-isomorphic irreducible representations of G. Choose bases $v_i \in V$, $i = 1 \ldots n$ and $w_a \in W$, $a = 1 \ldots m$. Then for any i, j, a, b, the matrix coefficients ρ_{ij}^V, ρ_{ab}^W are orthogonal: $(\rho_{ij}^V, \rho_{ab}^W) = 0$, where $(\,,\,)$ is the inner product on $C^\infty(G, \mathbb{C})$ given by*

$$(f_1, f_2) = \int_G f_1(g) \overline{f_2(g)} \, dg. \tag{4.1}$$

(2) *Let V be an irreducible representation of G and let $v_i \in V$ be an orthonormal basis with respect to a G-invariant inner product (which exists by Theorem 4.40). Then the matrix coefficients ρ_{ij}^V are pairwise orthogonal, and each has norm squared $1/\dim V$:*

$$(\rho_{ij}^V), \rho_{kl}^V = \frac{\delta_{ik} \delta_{jl}}{\dim V} \tag{4.2}$$

Proof. The proof is based on the following easy lemma.

Lemma 4.42.

(1) *Let V, W be non-isomorphic irreducible representations of G and f a linear map $V \to W$. Then $\int_G gfg^{-1} \, dg = 0$.*
(2) *If V is an irreducble representation and f is a linear map $V \to V$, then $\int g f g^{-1} \, dg = (\mathrm{tr}(f)/\dim V) \, \mathrm{id}$.*

Indeed, let $\tilde{f} = \int_G gfg^{-1} \, dg$. Then $\tilde{f}$ commutes with action of G: $h\tilde{f}h^{-1} = \int_G (hg)f(hg)^{-1} \, dg = \tilde{f}$. By Schur's lemma, $\tilde{f} = 0$ for $W \neq V$ and $\tilde{f} = \lambda \, \mathrm{id}$ for $W = V$. Since $\mathrm{tr}(gfg^{-1}) = \mathrm{tr} f$, we see that $\mathrm{tr} \tilde{f} = \mathrm{tr} f$, so $\lambda = \lambda(\mathrm{tr}(f)/\dim V)$. This proves the lemma.

Now let v_i, w_a be orthonormal bases in V, W with respect to a G-invariant inner product. Choose a pair of indices i, a and apply this lemma to the map $E_{ai} : V \to W$ given by $E_{ai}(v_i) = w_a$, $E_{ai}v_j = 0$, $j \neq i$. Then we have

$$\int_G \rho^W(g) E_{ai} \rho^V(g^{-1}) \, dg = 0.$$

Rewriting this in the matrix form and using $\rho(g^{-1}) = \overline{\rho(g)^t}$ (which follows from unitarity of $\rho(g)$), we get that for any b, j,

$$\int \rho_{ba}^W(g) \overline{\rho_{ji}^V(g)} \, dg = 0,$$

which proves the first part of the theorem in the case when the bases are orthonormal; the general case immediately follows.

To prove the second part, apply the lemma to a matrix unit $E_{ki} : V \to V$ to get

$$\sum_{l,j} E_{lj} \int \rho_{lk}^V(g)\overline{\rho_{ji}^V(g)} \, dg = \frac{\mathrm{tr}\, E_{ki}}{\dim V}\, \mathrm{id},$$

which immediately yields the second part of the theorem. □

So irreducible representations give us a way of constructing an orthonormal set of functions on the group. Unfortunately, they depend on the choice of basis. However, there is one particular combination of matrix coefficients that does not depend on the choice of basis.

Definition 4.43. A character of a representation V is the function on the group defined by

$$\chi_V(g) = \mathrm{tr}_V\, \rho(g) = \sum \rho_{ii}^V(g).$$

It is immediate from the definition that the character does not depend on the choice of basis in V. It also has a number of other properties, listed below; proof of them is left to the reader as an exercise.

Lemma 4.44.

(1) *Let $V = \mathbb{C}$ be the trivial representation. Then $\chi_V = 1$.*
(2) $\chi_{V \oplus W} = \chi_V + \chi_W$.
(3) $\chi_{V \otimes W} = \chi_V \chi_W$
(4) $\chi_V(ghg^{-1}) = \chi_V(h)$.
(5) *Let V^* be the dual of representation V. Then $\chi_{V^*} = \overline{\chi_V}$.*

The orthogonality relation for matrix coefficients immediately implies the following result for the characters.

Theorem 4.45.

(1) *Let V, W be non-isomorphic complex irreducible representations of a compact real Lie group G. Then the characters χ_V, χ_W are orthogonal with respect to inner product (4.1): $(\chi_V, \chi_W) = 0$.*
(2) *For any irreducible representation V, $(\chi_V, \chi_V) = 1$.*

In other words, if we denote by $\widehat{G}$ the set of isomorphism classes of irreducible representations of G, then the set $\{\chi_V, V \in \widehat{G}\}$ is an orthonormal family of functions on G.

This immediately implies a number of corollaries.

Corollary 4.46. *Let V be a complex representation of a compact real Lie group G. Then*

(1) *V is irreducible iff* $(\chi_V, \chi_V) = 1$.
(2) *V can be uniquely written in the form* $V \simeq \bigoplus n_i V_i$, V_i *– pairwise non-isomorphic irreducible representations, and the multiplicities* n_i *are given by* $n_i = (\chi_V, \chi_{V_i})$.

In principle, this theorem gives a way of computing multiplicites n_i. In real life, it is only usable for finite groups and some special cases. Much more practical ways of finding multiplicities will be given later when we develop weight decomposition for representations of semisimple Lie algebras (see Section 8.6).

Finally, let us return to the matrix coefficients of representations. One might ask whether it is possible to give a formulation of Theorem 4.41 in a way that does not require a choice of basis. The answer is "yes". Indeed, let $v \in V$, $v^* \in V^*$. Then we can define a function on the group $\rho_{v^*,v}(g)$ by

$$\rho_{v^*,v}(g) = \langle v^*, \rho(g)v \rangle.$$

This is a generalization of a matrix coefficient: if $v = v_j$, $v^* = v_i^*$, we recover matrix coefficient $\rho_{ij}(g)$.

This shows that for any representation V, we have a map

$$m \colon V^* \otimes V \to C^\infty(G, \mathbb{C})$$
$$v^* \otimes v \mapsto \langle v^*, \rho(g)v \rangle.$$

The space $V^* \otimes V$ has additional structure. First, we have two commuting actions of G on it, given by action on the first factor and on the second one; in other words, $V^* \otimes V$ is a G-bimodule. In addition, if V is unitary, then the inner product defines an inner product on V^* (the simplest way to define it is to say that if v_i is an orthonormal basis in V, then the dual basis v_i^* is an orthonormal basis in V^*). Define an inner product on $V^* \otimes V$ by

$$(v_1^* \otimes w_1, v_2^* \otimes w_2) = \frac{1}{\dim V}(v_1^*, v_2^*)(w_1, w_2). \tag{4.3}$$

Theorem 4.47. *Let* $\widehat{G}$ *be the set of isomorphism classes of irreducible representations of G. Define the map*

$$m \colon \bigoplus_{V_i \in \widehat{G}} V_i^* \otimes V_i \to C^\infty(G, \mathbb{C}) \tag{4.4}$$

by $m(v^* \otimes v)(g) = \langle v^*, \rho(g)v \rangle$. (*Here $\bigoplus$ is the algebraic direct sum, i.e. the space of finite linear combinations.*) *Then*

(1) *The map m is a morphism of G-bimodules:*

$$m((gv^*) \otimes v) = L_g(m(v^* \otimes v))$$
$$m(v^* \otimes gv) = R_g(m(v^* \otimes v)),$$

where L_g, R_g are the left and right actions of G on $C^\infty(G, \mathbb{C})$: $(L_g f)(h) = f(g^{-1}h)$, $(R_g f)(h) = f(hg)$.

(2) *The map m preserves the inner product, if we define the inner product in $\bigoplus_i V_i^* \otimes V_i$ by (4.3) and inner product in $C^\infty(G)$ by (4.1).*

Proof. The first part is obtained by explicit computation:

$$(R_g m(v^* \otimes v))(h) = m(v^* \otimes v)(hg) = \langle v^*, \rho(hg)v \rangle$$
$$= \langle v^*, \rho(h)\rho(g)v \rangle = m(v^* \otimes gv)(h)$$
$$(L_g m(v^* \otimes v))(h) = m(v^* \otimes v)(g^{-1}h) = \langle v^*, \rho(g^{-1})\rho(h)v \rangle$$
$$= \langle gv^*, \rho(h)v \rangle = m(gv^* \otimes v)(h)$$

The second part immediately follows from Theorem 4.41. □

Corollary 4.48. *The map m is injective.*

It turns out that this map is also surjective if we replace the algebraic direct sum by an appropriate completion: every function on the group can be approximated by a linear combination of matrix coefficients. The precise statement is known as Peter–Weyl theorem.

Theorem 4.49. *The map (4.4) gives an isomorphism*

$$\widehat{\bigoplus}_{V_i \in \hat{G}} V_i^* \otimes V_i \to L^2(G, \mathrm{d}g)$$

where $\widehat{\bigoplus}$ is the Hilbert space direct sum, i.e. the completion of the algebraic direct sum with respect to the metric given by inner product (4.3), and $L^2(G, \mathrm{d}g)$ is the Hilbert space of complex-valued square-integrable functions on G with respect to the Haar measure, with the inner product defined by (4.1).

The proof of this theorem requires some non-trivial analytic considerations and goes beyond the scope of this book. The interested reader can find it in [48] or [32].

Corollary 4.50. *The set of characters* $\{\chi_V, V \in \widehat{G}\}$ *is an orthonormal basis (in the sense of Hilbert spaces) of the space* $(L^2(G, \mathrm{d}g))^G$ *of conjugation-invariant functions on G.*

Example 4.51. Let $G = S^1 = \mathbb{R}/\mathbb{Z}$. As we have already discussed, the Haar measure on G is given by $\mathrm{d}x$ and the irreducible representations are parametrized by $\mathbb{Z}$: for any $k \in \mathbb{Z}$, we have one-dimensional representation V_k with the action of S^1 given by $\rho(a) = \mathrm{e}^{2\pi i k a}$ (see Example 4.27). The corresponding matrix coefficient is the same as character and is given by $\chi_k(a) = \mathrm{e}^{2\pi i k a}$.

Then the orthogonality relation of Theorem 4.41 gives

$$\int_0^1 \mathrm{e}^{2\pi i k x}\overline{\mathrm{e}^{2\pi i l x}}\,\mathrm{d}x = \delta_{kl},$$

which is the usual orthogonality relation for exponents. The Peter–Weyl theorem in this case just says that the exponents $\mathrm{e}^{2\pi i k x}, k \in \mathbb{Z}$, form an orthonormal basis of $L^2(S^1, \mathrm{d}x)$ which is one of the main statements of the theory of Fourier series: every L^2 function on S^1 can be written as a series $f(x) = \sum_{k \in \mathbb{Z}} c_k \mathrm{e}^{2\pi i k x}$ which converges in L^2 metric. For this reason, the study of the structure of $L^2(G)$ can be considered as a generalization of harmonic analysis.

4.8. Representations of $\mathfrak{sl}(2, \mathbb{C})$

In this section we will give a complete description of the representation theory of the Lie algebra $\mathfrak{sl}(2, \mathbb{C})$. This an instructive example; moreover, these results will be used as a basis for analysis of more complicated Lie algebras later.

Throughout this section, all representations are complex and finite-dimensional unless specified otherwise. For brevity, for a vector v in a representation V and $x \in \mathfrak{sl}(2, \mathbb{C})$, we will write xv instead of more accurate but cumbersome notation $\rho(x)v$.

Theorem 4.52. *Any representation of* $\mathfrak{sl}(2, \mathbb{C})$ *is completely reducible.*

Proof. By Lemma 4.4, representations of $\mathfrak{sl}(2, \mathbb{C})$ are the same as representations of $\mathfrak{su}(2)$ which in turn are the same as representations of $SU(2)$. Since the group $SU(2)$ is compact, by Theorem 4.40, every representation is completely reducible. $\square$

Remark 4.53. It is also possible to give a purely algebraic proof of complete reducibilty; one such proof, in a much more general situation, will be given later in Section 6.3.

Thus, our primary goal will be the classification of irreducible representations.

Recall that $\mathfrak{sl}(2, \mathbb{C})$ has a basis e, f, h with the commutation relations

$$[e,f] = h, \qquad [h,e] = 2e, \qquad [h,f] = -2f \tag{4.5}$$

(see Section 3.10). As will be proved later, this Lie algebra is simple (Example 5.38).

The main idea of the study of representations of $\mathfrak{sl}(2, \mathbb{C})$ is to start by diagonalizing the operator h.

Definition 4.54. Let V be a representation of $\mathfrak{sl}(2, \mathbb{C})$. A vector $v \in V$ is called vector of weight λ, $\lambda \in \mathbb{C}$, if it is an eigenvector for h with eigenvalue λ:

$$hv = \lambda v.$$

We denote by $V[\lambda] \subset V$ the subspace of vectors of weight λ.

The following lemma plays the key role in the study of representations of $\mathfrak{sl}(2, \mathbb{C})$.

Lemma 4.55.

$$eV[\lambda] \subset V[\lambda + 2]$$
$$fV[\lambda] \subset V[\lambda - 2].$$

Proof. Let $v \in V[\lambda]$. Then

$$hev = [h, e]v + ehv = 2ev + \lambda ev = (\lambda + 2)ev$$

so $ev \in V[\lambda + 2]$. The proof for f is similar. $\qquad\square$

Theorem 4.56. *Every finite-dimensional representation V of* $\mathfrak{sl}(2, \mathbb{C})$ *can be written in the form*

$$V = \bigoplus_{\lambda} V[\lambda]$$

where $V[\lambda]$ is defined in Definition 4.54. This decomposition is called the weight decomposition of V.

Proof. Since every representation of $\mathfrak{sl}(2, \mathbb{C})$ is completely reducible, it suffices to prove this for irreducible V. So assume that V is irreducible. Let $V' = \sum_{\lambda} V[\lambda]$ be the subspace spanned by eigenvectors of h. By well-known result of linear algebra, eigenvectors with different eigenvalues are linearly independent, so $V' = \bigoplus V[\lambda]$. By Lemma 4.55, V' is stable under the action of e, f and h. Thus, V' is a subrepresentation. Since we assumed that V is irreducible, and $V' \neq 0$ (h has at least one eigenvector), we see that $V' = V$. $\square$

Our main goal will be classification of ireducible finite-dimensional representations. So from now on, let V be an irreducible representation of $\mathfrak{sl}(2, \mathbb{C})$.

Let λ be a weight of V (i.e., $V[\lambda] \neq 0$) which is maximal in the following sense:

$$\text{Re } \lambda \geq \text{Re } \lambda' \qquad \text{for every weight } \lambda' \text{ of } V. \tag{4.6}$$

Such a weight will be called a "highest weight of V", and vectors $v \in V[\lambda]$ will be called highest weight vectors. It is obvious that every finite-dimensional representation has at least one non-zero highest weight vector.

Lemma 4.57. *Let $v \in V[\lambda]$ be a highest weight vector in V.*

(1) $ev = 0$.
(2) *Let*

$$v^k = \frac{f^k}{k!} v, \qquad k \geq 0$$

Then we have

$$hv^k = (\lambda - 2k)v^k,$$
$$fv^k = (k+1)v^{k+1}, \tag{4.7}$$
$$ev^k = (\lambda - k + 1)v^{k-1}, \quad k > 0$$

Proof. By Lemma 4.55, $ev \in V[\lambda + 2]$. But by definition of a highest weight vector, $V[\lambda + 2] = 0$. This proves the first part.

To prove the second part, note that the formula for the action of f is immediate from the definition, and formula for the action of h follows from Lemma 4.55. Thus, we need to prove the formula for the action of e.

The proof goes by induction. For $k = 1$ we have

$$ev^1 = efv = [e, f]v + fev = hv = \lambda v$$

(using $ev = 0$).

The induction step is proved by

$$ev^{k+1} = \frac{1}{k+1}ef\,v^k = \frac{1}{k+1}(hv^k + fev^k)$$

$$= \frac{1}{k+1}\left((\lambda - 2k)v^k + (\lambda - k + 1)fv^{k-1}\right)$$

$$= \frac{1}{k+1}(\lambda - 2k + (\lambda - k + 1)k)v^k = (\lambda - k)v^k.$$

□

Of course, since V is finite-dimensional, only finitely many of v^k are non-zero. However, it is convenient to consider V as a quotient of infinite-dimensional vector space with basis v^k. This is done as follows.

Lemma 4.58. *Let* $\lambda \in \mathbb{C}$. *Define* M_λ *to be the infinite-dimensional vector space with basis* $v^0, v^1, \ldots$.

(1) *Formulas (4.7) and* $ev^0 = 0$ *define on* M_λ *the structure of an (infinite-dimensional) representation of* $\mathfrak{sl}(2,\mathbb{C})$.

(2) *If* V *is an irreducible finite-dimensional representation of* $\mathfrak{sl}(2,\mathbb{C})$ *which contains a non-zero highest weight vector of highest weight* λ, *then* $V = M_\lambda/W$ *for some subrepresentation* W.

Proof. The first part is done by explicit calculation which is essentially equivalent to the calculation used in the proof of Lemma 4.57. The second part immediately follows from Lemma 4.57. □

Now we can prove the main theorem.

Theorem 4.59.

(1) *For any* $n \geq 0$, *let* V_n *be the finite-dimensional vector space with basis* $v^0, v^1, \ldots, v^n$. *Define the action of* $\mathfrak{sl}(2,\mathbb{C})$ *by*

$$hv^k = (n - 2k)v^k,$$

$$fv^k = (k+1)v^{k+1}, \quad k < n; \quad fv^n = 0 \tag{4.8}$$

$$ev^k = (n + 1 - k)v^{k-1}, \quad k > 0; \quad ev^0 = 0.$$

Then V_n *is an irreducible representation of* $\mathfrak{sl}(2,\mathbb{C})$; *we will call it the irreducible representation with highest weight* n.

(2) *For* $n \neq m$, *representation* V_n, V_m *are non-isomorphic.*

(3) *Every finite-dimensional irreducible representation of* $\mathfrak{sl}(2,\mathbb{C})$ *is isomorphic to one of representations* V_n.

Proof. Consider the infinite-dimensional representation M_λ defined in Lemma 4.58. If $\lambda = n$ is a non-negative integer, consider the subspace $M' \subset M_n$ spanned by vectors $v^{n+1}, v^{n+2}, \ldots$. Then this subspace is actually a subrepresentation. Indeed, it is obviously stable under the action of h and f; the only non-trivial relation to check is that $ev^{n+1} \subset M'$. But $ev^{n+1} = (n + 1 - (n + 1))v^n = 0$.

Thus, the quotient space M_n/M' is a finite-dimensional representation of $\mathfrak{sl}(2, \mathbb{C})$. It is obvious that it has basis $v^0, \ldots, v^n$ and that the action of $\mathfrak{sl}(2, \mathbb{C})$ is given by (4.8). Irreducibility of this representation is also easy to prove: any subrepresentation must be spanned by some subset of $v, v^1, \ldots, v^n$, but it is easy to see that each of them generates (under the action of $\mathfrak{sl}(2, \mathbb{C})$) the whole representation V_n. Therefore, V_n is an irreducible finite-dimensional representation of $\mathfrak{sl}(2, \mathbb{C})$. Sonce $\dim V_n = n + 1$, it is obvious that V_n are pairwise non-isomorphic.

To prove that every irreducible representation is of this form, let V be an irreducible representation of $\mathfrak{sl}(2, \mathbb{C})$ and let $v \in V[\lambda]$ be a highest weight vector. By Lemma 4.58, V is a quotient of M_λ; in other words, it is spanned by vectors $v^k = (f^k/k!)v$.

Since v^k have different weights, if they are non-zero, then they must be linearly independent. On the other hand, V is finite-dimensional; thus, only finitely many of v^i are non-zero. Let n be maximal such that $v^n \neq 0$, so that $v^{n+1} = 0$. Obviously, in this case $v^0, \ldots, v^n$ are all non-zero and since they have different weight, they are linearly independent, so they form a basis in V.

Since $v^{n+1} = 0$, we must have $ev^{n+1} = 0$. On the other hand, by (4.7), we have $ev^{n+1} = (\lambda - n)v^n$. Since $v^n \neq 0$, this implies that $\lambda = n$ is a non-negative integer. Thus, V is a representation of the form discussed in part (1). $\square$

Figure 4.1 illustrates action of $\mathfrak{sl}(2, \mathbb{C})$ in V_n.

Irreducible representations V_n can also be described more explicitly, as symmetric powers of the usual two-dimensional representation (see Exercise 4.12).

As a corollary, we immediately get some useful information about any finite-dimensional representation of $\mathfrak{sl}(2, \mathbb{C})$.

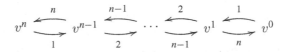

Figure 4.1 Action of $\mathfrak{sl}(2, \mathbb{C})$ in the irreducible representation V_n. Top arrows show the action of f, bottom arrows show the action of e.

Theorem 4.60. *Let* V *be a finite-dimensional complex representation of* $\mathfrak{sl}(2, \mathbb{C})$.

(1) V *admits a weight decomposition with integer weights:*

$$V = \bigoplus_{n \in \mathbb{Z}} V[n].$$

(2) $\dim V[n] = \dim V[-n]$. *Moreover, for* $n \geq 0$ *the maps*

$$e^n : V[n] \to V[-n]$$
$$f^n : V[-n] \to V[n]$$

are isomorphisms.

Proof. Since every representation is completely reducible, it suffices to prove this in the case when $V = V_n$ is an irreducible representation. In this case, it follows from Theorem 4.59. $\qquad\square$

By results of Section 4.1, this also implies similar statements for representations of Lie algebra $\mathfrak{so}(3, \mathbb{R})$ and the group $SO(3, \mathbb{R})$. These results are given in Exercise 4.13.

4.9. Spherical Laplace operator and the hydrogen atom

In this section, we apply our knowledge of representation theory of Lie groups and Lie algebras to the study of the Laplace operator on the sphere, thus answering the question raised in the introduction. The material of this section will not be used in the rest of the book, so it can be safely skipped. However, it is a very illustrative example of how one uses representation theory in the study of systems with a symmetry.

Let $\Delta = \partial_x^2 + \partial_y^2 + \partial_z^2$ be the usual Laplace operator in $\mathbb{R}^3$. We would like to split it into "radial" and "spherical" parts, which can be done as follows.

Notice that $\mathbb{R}^3 - \{0\}$ can be identified with the direct product

$$\mathbb{R}^3 - \{0\} \simeq S^2 \times \mathbb{R}_+$$
$$\vec{x} \mapsto (\mathbf{u}, r) \tag{4.9}$$
$$\mathbf{u} = \frac{\vec{x}}{|\vec{x}|} \in S^2, \quad r = |\vec{x}| \in \mathbb{R}_+.$$

The following well-known lemma shows how Δ can be rewritten in coordinates $\mathbf{u}, r$.

Lemma 4.61.

(1) *When rewritten in coordinates* **u**, *r, we have*

$$\Delta = \frac{1}{r^2}\Delta_{\text{sph}} + \Delta_{\text{radial}},$$

where Δ_{sph} *is a differential operator on* S^2 *and* Δ_{radial} *is a differential operator on* $\mathbb{R}_+$.

(2) *We have*

$$\Delta_{\text{radial}} = \partial_r^2 + \frac{2}{r}\partial_r \tag{4.10}$$

$$\Delta_{\text{sph}} = J_x^2 + J_y^2 + J_z^2,$$

where

$$J_x = y\partial_z - z\partial y$$
$$J_y = z\partial_x - x\partial z$$
$$J_z = x\partial_y - y\partial x$$

are vector fields corresponding to the generators of Lie algebra $\mathfrak{so}(3,\mathbb{R})$ *(see Exercise 3.11).*

Sketch of proof. Since for any $r > 0$, the vector fields J_x, J_y, J_z are tangent to the sphere of radius r, the operator Δ_{sph} defined by (4.10) is well defined as a differential operator on the sphere. Identity $\Delta = (1/r^2)\Delta_{\text{sph}} + \Delta_{\text{radial}}$ can be shown by explicit calculation (see Exercise 3.11). □

One can introduce the usual coordinates on the sphere and write Δ_{sph} in these coordinates. Such an expression can be found in any book on multivariable calculus, but it is very messy; more importantly, it will be useless for our purposes. For this reason it is not given here.

The main question we want to answer is as follows:

Find eigenvalues of Δ_{sph} acting on functions on S^2 (4.11)

The motivation for this problem comes from physics. Namely, the quantum mechanical description of a particle moving in a central force field (for example, an electron in the hydrogen atom) is given by Schrödinger equation

$$\dot{\psi} = iH\psi$$

where $\psi = \psi(t, \vec{x})$, $\vec{x} \in \mathbb{R}^3$, is the wave-function which describes the state of the system and

$$H = -\Delta + V(r)$$

is the Hamiltonian, or the energy operator; here $V(r)$ is the potential, which describes the central force field. Solving the Schrödinger equation is essentially equivalent to diagonalizing the Hamiltonian. The usual approach to this problem is to use separation of variables, writing

$$\psi(\vec{x}) = \sum f_i(r) g_i(\mathbf{u}) \tag{4.12}$$

where $r \in \mathbb{R}_+$, $\mathbf{u} \in S^2$ are given by (4.9), and g_i are eigenfunctions for Δ_{sph}. Substituting this in the equation $H\psi = \lambda\psi$ gives a second-order differential equation on $f_i(r)$. For many potentials $V(r)$, one can explicitly solve this equation, thus giving eigenfunctions of the Hamiltonian – in particular, the energy levels for the hydrogen atom. Details can be found, for example, in [34].

Returning to question (4.11), we notice that the straightforward approach, based on introducing coordinates on the sphere and writing the corresponding partial differential equation, is rather complicated. Instead, we will use the symmetries of the sphere, as was outlined in the introduction. We have an obvious action of the group $G = SO(3, \mathbb{R})$ on the sphere S^2 which therefore defines an action of G on the space of functions on S^2, by $g.f(x) = f(g^{-1}(x))$.

Lemma 4.62. $\Delta_{\text{sph}} \colon C^\infty(S^2) \to C^\infty(S^2)$ *commutes with the action of* $SO(3, \mathbb{R})$.

Proof. This can be shown in several ways. The easiest way is to note that it is well known that the Laplace operator Δ is rotation invariant. Obviously, the radial part Δ_{radial} is also rotation invariant; thus, $\Delta_{\text{sph}} = r^2(\Delta - \Delta_{\text{radial}})$ is also rotation invariant.

An alternative way of showing the rotational invariance of Δ_{sph} is by using equality $\Delta_{\text{sph}} = J_x^2 + J_y^2 + J_z^2$. Indeed, it suffices to show that in any representation V of $SO(3, \mathbb{R})$, the operator $C = \rho(J_x)^2 + \rho(J_y)^2 + \rho(J_z)^2$ commutes with the action of $SO(3, \mathbb{R})$. By the results of Section 4.2, it is equivalent to checking that for any $a \in \mathfrak{g}$, we have $[\rho(a), C] = 0$. This can be easily shown by explicit calculation. We do not give this computation here, as in the next chapter we will develop the theory of universal enveloping algebras which provides a natural language for computations of this sort; in particular, we will show that C is a central element in the universal enveloping algebra $U\mathfrak{so}(3, \mathbb{R})$. Later we will show that the fact that C is central can be derived from general theory of Casimir elements, thus making the above computation unnecessary. See Exercise 6.1 for details.   □

Therefore, by the general results of Section 4.4, the best way to study Δ_{sph} would be to decompose the space of functions on S^2 into irreducible representations of $SO(3, \mathbb{R})$. As usual, it is more convenient to work with complex representations, so we consider the space of complex-valued functions.

There are some obvious technical problems: the space of functions is infinite dimensional. To avoid dealing with convergence questions and other analytical difficulties, let us consider the space of polynomials

$$P_n = \left\{ \begin{array}{l} \text{Complex-valued functions on } S^2 \text{ which can be written as} \\ \text{polynomials in } x, y, z \text{ of total degree} \leq n \end{array} \right\}.$$

(4.13)

One easily sees that each P_n is a finite-dimensional representation of $SO(3, \mathbb{R})$ which is also Δ_{sph}-invariant. Thus, we can use the theory of finite-dimensional representations to decompose P_n into irreducible representations and then use this to find the eigenvalues of Δ_{sph} in P_n. Since $\bigcup P_n = P$ is the space of all polynomial functions on S^2, which is everywhere dense in $C^\infty(S^2)$, diagonalizing Δ_{sph} in P is essentially equivalent to diagonalizing Δ_{sph} in C^∞ (the precise statement will be given below).

Thus, our immediate goal is to decompose P_n into a direct sum of irreducible representations of $SO(3, \mathbb{R})$. To do this, note that by results of Exercise 4.13, irreducible representations of $SO(3, \mathbb{R})$ are of the form $V_{2k}, k \in \mathbb{Z}_+$. Thus, we can write

$$P_n = \bigoplus c_k V_{2k}.$$

To find coefficients c_k, we need to find the character of P_n, i.e., the dimensions of eigenspaces for J_z (recall that under the isomorphism $\mathfrak{so}(3, \mathbb{R})_{\mathbb{C}} \simeq \mathfrak{sl}(2, \mathbb{C})$ constructed in Section 3.10, J_z is identified with $ih/2$). We can do it by explicitly constructing an eigenbasis in P_n.

Lemma 4.63. *The following set of functions form a basis of P_n:*

$$f_{p,k} = z^p \left(\sqrt{1 - z^2} \right)^{|k|} e^{ik\varphi}, \qquad p \in \mathbb{Z}_+, k \in \mathbb{Z}, p + |k| \leq n,$$

where φ is defined by $x = \rho \cos \varphi, y = \rho \sin \varphi, \rho = \sqrt{x^2 + y^2}$.

Proof. Let $u = x + iy = \rho e^{i\varphi}$, $v = x - iy = \rho e^{-i\varphi}$. Then any polynomial in x, y, z can be written as a polynomial in z, u, v. Since on the sphere we have

$1 - z^2 = x^2 + y^2 = uv$, every monomial $z^k u^l v^m$ can be written as a monomial which involves only u or v but not both. Thus, every element of P_n can be written as a linear combination of monomials

$$z^p,$$

$$z^p u^k = z^p \rho^k e^{ik\varphi} = f_{p,k},$$

$$z^p v^k = z^p \rho^k e^{-ik\varphi} = f_{p,-k}$$

with $p, k \in \mathbb{Z}_+, p + k \leq n$. Thus, elements $f_{p,k}$ span P_n.

To show that they are linearly independent, assume that

$$\sum_{k} a_{p,k} f_{p,k} = \sum_{k} a_k(z) e^{ik\varphi} = 0, \qquad a_k(z) = \sum_{p} a_{p,k} z^p (\sqrt{1 - z^2})^{|k|}.$$

By the uniqueness of Fourier series, we see that for every $k \in \mathbb{Z}$, $z \in (-1, 1)$, we have $a_k(z) = 0$ which easily implies that for every p, k, $a_{p,k} = 0$. $\square$

We can now find the dimensions of the eigenspaces for J_z. Since J_z is the generator of rotations around z axis, it is easy to see that in the cylindrical coordinates z, ρ, φ, $J_z = \frac{\partial}{\partial \varphi}$. Thus,

$$J_z f_{p,k} = ik f_{p,k}$$

so $P_n[2k] = \mathrm{Span}(f_{p,k})_{0 \leq p \leq n - |k|}$ and thus $\dim P_n[2k] = n + 1 - k$. Using the formula for multiplicities from Exercise 4.11, we see that

$$P_n \simeq V_0 \oplus V_2 \oplus \cdots \oplus V_{2n}. \tag{4.14}$$

Now the computation of the eigenvalues of spherical Laplace operator is easy. Namely, by Exercise 4.4, $J_x^2 + J_y^2 + J_z^2$ acts in V_l by $-l(l+2)/4$. Thus, we get the following result.

Theorem 4.64. *The eigenvalues of the spherical Laplace operator Δ_{sph} in the space P_n are*

$$\lambda_k = -k(k+1), \qquad k = 0, \ldots, n \tag{4.15}$$

and multiplicity of λ_k is equal to $\dim V_{2k} = 2k + 1$.

Finally, we can formulate the final result about eigenfunctions in $C^\infty(S^2)$.

Theorem 4.65. *Each eigenfunction of Δ_{sph} is polynomial. The eigenvalues are given by (4.15), and multiplicity of λ_k is equal to $2k + 1$.*

Proof. Consider the space $L^2(S^2, \mathbb{C})$ of complex-valued L^2 functions on S^2. Since action of SO(3) preserves the volume form, it also preserves the inner product in $L^2(S^2, \mathbb{C})$. It shows that operators J_x, J_y, J_z are skew-Hermitian, and thus, Δ_{sph} is Hermitian, or self-adjoint.

Let $E_n \subset P_n$ be the orthogonal complement to P_{n-1}. Then E_n is SO(3)-invariant, and it follows from (4.14) that as an SO(3)-module $E_n \simeq V_{2n}$, so Δ_{sph} acts on E_n by λ_n. On the other hand, since the space of polynomials is dense in L^2, we have

$$L^2(S^2, \mathbb{C}) = \bigoplus_{n \geq 0} E_n$$

(direct sum of Hilbert spaces). Thus, if $\Delta_{\text{sph}} f = \lambda f$ for some function $f \in C^\infty(S^2) \subset L^2(S^2)$, then either $\lambda \neq \lambda_n$ for all n, which forces $(f, E_n) = 0$ for all n, so $f = 0$, or $\lambda = \lambda_n$, so $(f, E_k) = 0$ for all $k \neq n$, so $f \in E_n$. $\square$

4.10. Exercises

4.1. Let $\varphi : SU(2) \to SO(3, \mathbb{R})$ be the covering map constructed in Exercise 2.8.

(1) Show that $\text{Ker}\, \varphi = \{1, -1\} = \{1, e^{\pi i h}\}$, where h is defined by (3.23).

(2) Using this, show that representations of $SO(3, \mathbb{R})$ are the same as representations of $\mathfrak{sl}(2, \mathbb{C})$ satisfying $e^{\pi i \rho(h)} = \text{id}$.

4.2. Let $V = \mathbb{C}^2$ be the standard two-dimensional representation of the Lie algebra $\mathfrak{sl}(2, \mathbb{C})$, and let $S^k V$ be the symmetric power of V.

(1) Write explicitly the action of $e, f, h \in \mathfrak{sl}(2, \mathbb{C})$ (see Section 3.10) in the basis $e_1^i e_2^{k-i}$.

(2) Show that $S^2 V$ is isomorphic to the adjoint representation of $\mathfrak{sl}(2, \mathbb{C})$.

(3) By results of Section 4.1, each representation of $\mathfrak{sl}(2, \mathbb{C})$ can be considered as a representation of $\mathfrak{so}(3, \mathbb{R})$. Which of representations $S^k V$ can be lifted to a representation of $SO(3, \mathbb{R})$?

4.3. Show that $\Lambda^n \mathbb{C}^n \simeq \mathbb{C}$ as a representation of $\mathfrak{sl}(n, \mathbb{C})$. Does it also work for $\mathfrak{gl}(n, \mathbb{C})$?

4.4. Let V be a representation of $\mathfrak{sl}(2, \mathbb{C})$, and let $C \in \text{End}(V)$ be defined by

$$C = \rho(e)\rho(f) + \rho(f)\rho(e) + \frac{1}{2}\rho(h)^2.$$

(1) Show that C commutes with the action of $\mathfrak{sl}(2,\mathbb{C})$: for any $x \in \mathfrak{sl}(2,\mathbb{C})$, we have $[\rho(x), C] = 0$. [Hint: use that for any a, b, $c \in \mathrm{End}(V)$, one has $[a, bc] = [a, b]c + b[a, c]$.]

(2) Show that if $V = V_k$ is an irreducible representation with highest weight k, then C is a scalar operator: $C = c_k \, \mathrm{id}$. Compute the constant c_k.

(3) Recall that we have an isomorphism $\mathfrak{so}(3,\mathbb{C}) \simeq \mathfrak{sl}(2,\mathbb{C})$ (see Section 3.10). Show that this isomorphism identifies operator C above with a multiple of $\rho(J_x)^2 + \rho(J_y)^2 + \rho(J_z)^2$.

The element C introduced here is a special case of more general notion of Casimir element which will be discussed in Section 6.3.

4.5. (1) Let V, W be irreducible representations of a Lie group G. Show that $(V \otimes W^*)^G = 0$ if V is non-isomorphic to W, and that $(V \otimes V^*)^G$ is canonically isomorphic to $\mathbb{C}$.

(2) Let V be an irreducible representation of a Lie algebra $\mathfrak{g}$. Show that V^* is also irreducible, and deduce from this that the space of $\mathfrak{g}$-invariant bilinear forms on V is either zero or 1-dimensional.

4.6. For a representation V of a Lie algebra $\mathfrak{g}$, define the space of *coinvaraints* by $V_{\mathfrak{g}} = V/\mathfrak{g}V$, where $\mathfrak{g}V$ is the subspace spanned by $xv, x \in \mathfrak{g}, v \in V$.

(1) Show that if V is completely reducible, then the composition $V^{\mathfrak{g}} \hookrightarrow V \to V_{\mathfrak{g}}$ is an isomorphism.

(2) Show that in general, it is not so. (Hint: take $\mathfrak{g} = \mathbb{R}$ and an appropriate representation V.)

4.7. Let $\mathfrak{g}$ be a Lie algebra, and $(\ ,\ )$ – a symmetric ad-invariant bilinear form on $\mathfrak{g}$. Show that the element $\omega \in (\mathfrak{g}^*)^{\otimes 3}$ given by

$$\omega(x, y, z) = ([x, y], z)$$

is skew-symmetric and ad-invariant.

4.8. Prove that if $A \colon \mathbb{C}^n \to \mathbb{C}^n$ is an operator of finite order: $A^k = I$ for some k, then A is diagonalizable. [Hint: use theorem about complete reducibility of representations of a finite group]

4.9. Let C be the standard cube in $\mathbb{R}^3$: $C = \{|x_i| \leq 1\}$, and let S be the set of faces of C (thus, S consists of six elements). Consider the six-dimensional complex vector V space of functions on S, and define $A \colon V \to V$ by

$$(Af)(\sigma) = \frac{1}{4} \sum_{\sigma'} f(\sigma')$$

where the sum is taken over all faces σ' which are neighbors of σ (i.e., have a common edge with σ). The goal of this problem is to diagonalize A.

(1) Let $G = \{g \in O(3, \mathbb{R}) \mid g(C) = C\}$ be the group of symmetries of C. Show that A commutes with the natural action of G on V.

(2) Let $z = -I \in G$. Show that as a representation of G, V can be decomposed in the direct sum

$$V = V_+ \oplus V_-, \qquad V_\pm = \{f \in V \mid zf = \pm f\}.$$

(3) Show that as a representation of G, V_+ can be decomposed in the direct sum

$$V_+ = V_+^0 \oplus V_+^1, \quad V_+^0 = \{f \in V_+ \mid \sum_\sigma f(\sigma) = 0\}, \quad V_+^1 = \mathbb{C} \cdot 1,$$

where 1 denotes the constant function on S whose value at every $\sigma \in S$ is 1.

(4) Find the eigenvalues of A on V_-, V_+^0, V_+^1.
[Note: in fact, each of V_-, V_+^0, V_+^1 is an irreducible representation of G, but you do not need this fact.]

4.10. Let $G = SU(2)$. Recall that we have a diffeomorphism $G \simeq S^3$ (see Example 2.5).

(1) Show that the left action of G on $G \simeq S^3 \subset \mathbb{R}^4$ can be extended to an action of G by linear orthogonal transformations on $\mathbb{R}^4$.

(2) Let $\omega \in \Omega^3(G)$ be a left-invariant 3-form whose value at $1 \in G$ is defined by

$$\omega(x_1, x_2, x_3) = \mathrm{tr}([x_1, x_2]x_3), \qquad x_i \in \mathfrak{g}$$

(see Exercise 4.7). Show that $\omega = \pm 4dV$ where dV is the volume form on S^3 induced by the standard metric in $\mathbb{R}^4$ (hint: let x_1, x_2, x_3 be some orthonormal basis in $\mathfrak{su}(2)$ with respect to $\frac{1}{2}\mathrm{tr}(a\bar{b}^t)$). (Sign depends on the choice of orientation on S^3.)

(3) Show that $\left(\frac{1}{8\pi^2}\right)\omega$ is a bi-invariant form on G such that for appropriate choice of orientation on G, $\left(\frac{1}{8\pi^2}\right)\int_G \omega = 1$.

4.11. Show that if V is a finite-dimensional representation of $\mathfrak{sl}(2, \mathbb{C})$, then $V \simeq \bigoplus n_k V_k$, and $n_k = \dim V[k] - \dim V[k+2]$. Show also that $\sum n_{2k} = \dim V[0]$, $\sum n_{2k+1} = \dim V[1]$.

4.12. Show that the symmetric power representation $S^k \mathbb{C}^2$, considered in Exercise 4.2, is isomorphic to the irreducible representation V_k with highest weight k.

4.13. Prove an analog of Theorem 4.60 for complex representations of $\mathfrak{so}(3, \mathbb{R})$, namely,

(1) Every finite-dimensional representation of $\mathfrak{so}(3, \mathbb{R})$ admits a weight decomposition:

$$V = \bigoplus_{n \in \mathbb{Z}} V[n],$$

where $V[n] = \{v \in V \mid J_z v = \frac{in}{2} v\}$.

(2) A representation V of $\mathfrak{so}(3, \mathbb{R})$ can be lifted to a representation of $SO(3, \mathbb{R})$ iff all weights are even: $V[k] = 0$ for all odd k (cf. with Exercise 4.1).

In physical literature, the number $j = \text{weight}/2$ is called the *spin*; thus, instead of talking say, of representation with highest weight 3, physicicts would talk about spin 3/2 representation. In this language, we see that a representation V of $\mathfrak{so}(3, \mathbb{R})$ can be lifted to a representation of $SO(3, \mathbb{R})$ iff the spin is integer.

4.14. Complete the program sketched in Section 4.9 to find the eigenvalues and multiplicities of the operator

$$H = -\Delta - \frac{c}{r}, \qquad c > 0$$

in $L^2(\mathbb{R}^3, \mathbb{C})$ (this operator describes the hydrogen atom).

5

Structure theory of Lie algebras

In this section, we will start developing the structure theory of Lie algebras, with the goal of getting eventually the full classification for semisimple Lie algebras and their representations.

In this chapter, $\mathfrak{g}$ will always stand for a finite-dimensional Lie algebra over the ground field $\mathbb{K}$ which can be either $\mathbb{R}$ or $\mathbb{C}$ (most results will apply equally in both cases and in fact for any field of characteristic zero). We will not be using the theory of Lie groups.

5.1. Universal enveloping algebra

In a Lie algebra $\mathfrak{g}$, in general there is no multiplication: the products of the form $xy, x, y \in \mathfrak{g}$, are not defined. However, if we consider a representation $\rho \colon \mathfrak{g} \to \mathfrak{gl}(V)$, then the product $\rho(x)\rho(y)$ is well-defined in such a representation – and in fact, as we will see later, operators of this kind can be very useful in the study of representations. Moreover, the commutation relations in $\mathfrak{g}$ imply some relations on the operators of this form. For example, commutation relation $[e, f] = h$ in $\mathfrak{sl}(2, \mathbb{C})$ implies that in any representation of $\mathfrak{sl}(2, \mathbb{C})$ we have $\rho(e)\rho(f) - \rho(f)\rho(e) = \rho(h)$, or equivalently, $\rho(e)\rho(f) = \rho(h) + \rho(f)\rho(e)$. These relations do not depend on the choice of representation ρ.

Motivated by this, we define the "universal" associative algebra generated by products of operators of the form $\rho(x), x \in \mathfrak{g}$.

Definition 5.1. Let $\mathfrak{g}$ be a Lie algebra over a field $\mathbb{K}$. The universal enveloping algebra of $\mathfrak{g}$, denoted by $U\mathfrak{g}$, is the associative algebra with unit over $\mathbb{K}$ with generators $i(x), x \in \mathfrak{g}$, subject to relations $i(x + y) = i(x) + i(y), i(cx) = ci(x), c \in \mathbb{K}$, and

$$i(x)i(y) - i(y)i(x) = i([x, y]). \tag{5.1}$$

To simplify the notation, we (and everyone else) will usually write simply $x \in U\mathfrak{g}$ instead of $i(x)$. This will be justified later (see Corollary 5.13) when we show that the map $i \colon \mathfrak{g} \to U\mathfrak{g}$ is injective and thus $\mathfrak{g}$ can be considered as a subspace in $U\mathfrak{g}$.

If we dropped relation (5.1), we would get the associative algebra generated by elements $x \in \mathfrak{g}$ with no relations other than linearity and associativity. By definition, this is exactly the tensor algebra of $\mathfrak{g}$:

$$T\mathfrak{g} = \bigoplus_{n \geq 0} \mathfrak{g}^{\otimes n}. \tag{5.2}$$

Thus, one can alternatively describe the universal enveloping algebra as the quotient of the tensor algebra:

$$U\mathfrak{g} = T\mathfrak{g}/(xy - yx - [x,y]), \qquad x, y \in \mathfrak{g}. \tag{5.3}$$

Example 5.2. Let $\mathfrak{g}$ be a commutative Lie algebra. Then $U\mathfrak{g}$ is generated by elements $x \in \mathfrak{g}$ with relations $xy = yx$. In other words, $U\mathfrak{g} = S\mathfrak{g}$ is the symmetric alebra of $\mathfrak{g}$, which can also be described as the algebra of polynomial functions on $\mathfrak{g}^*$. Choosing a basis x_i in $\mathfrak{g}$ we see that $U\mathfrak{g} = S\mathfrak{g} = \mathbb{K}[x_1, \ldots, x_n]$.

Note that in this example, the universal enveloping algebra is infinite-dimensional. In fact, $U\mathfrak{g}$ is always infinite-dimensional (unless $\mathfrak{g} = 0$). We will return to the discussion of the size of $U\mathfrak{g}$ in the next section.

Example 5.3. The universal enveloping algebra of $\mathfrak{sl}(2, \mathbb{C})$ is the associative algebra over $\mathbb{C}$ generated by elements e, f, h with the relations $ef - fe = h$, $he - eh = 2e$, $hf - fh = -2f$.

It should be noted that even when $\mathfrak{g} \subset \mathfrak{gl}(n, \mathbb{K})$ is a matrix algebra, multiplication in $U\mathfrak{g}$ is different from multiplication of matrices. For example, let e be the standard generator of $\mathfrak{sl}(2, \mathbb{C})$. Then $e^2 = 0$ as a 2×2 matrix, but $e^2 \neq 0$ in $U\mathfrak{g}$ – and for a good reason: there are many representations of $\mathfrak{sl}(2, \mathbb{C})$ in which $\rho(e)^2 \neq 0$.

The following theorem shows that $U\mathfrak{g}$ is indeed universal in a certain sense, which justifies the name.

Theorem 5.4. *Let A be an associative algebra with unit over $\mathbb{K}$ and let $\rho \colon \mathfrak{g} \to A$ be a linear map such that $\rho(x)\rho(y) - \rho(y)\rho(x) = \rho([x,y])$. Then ρ can be uniquely extended to a morphism of associative algebras $U\mathfrak{g} \to A$.*

Corollary 5.5. *Any representation of $\mathfrak{g}$ (not necessarily finite-dimensional) has a canonical structure of a $U\mathfrak{g}$-module. Conversely, every $U\mathfrak{g}$-module has a canonical structure of a representation of $\mathfrak{g}$.*

In other words, categories of representations of $\mathfrak{g}$ and of $U\mathfrak{g}$-modules are equivalent.

As a useful application of this result, we can use $U\mathfrak{g}$ to construct various operators acting in representations of $\mathfrak{g}$ – in particular to construct intertwining operators.

Example 5.6. Let $C = ef + fe + \frac{1}{2}h^2 \in U\mathfrak{sl}(2, \mathbb{C})$. Then

$$eC = e^2f + efe + \frac{1}{2}eh^2 = e(fe + h) + (fe + h)e + \frac{1}{2}(he - 2e)h$$

$$= efe + fe^2 + \frac{1}{2}heh + eh + he - eh = efe + fe^2 + \frac{1}{2}h(he - 2e) + he$$

$$= efe + fe^2 + \frac{1}{2}h^2e = Ce.$$

The idea of this calculation is to move e to the right, using the relations $ef = fe + h$, $eh = he - 2e$ to interchange it with f, h. Similar calculations also show that $fC = Cf, hC = Ch$. Thus, C is central in $U\mathfrak{g}$.

In particular, this implies that in every representation V of $\mathfrak{sl}(2, \mathbb{C})$, the element $\rho(C) \colon V \to V$ commutes with the action of $\mathfrak{sl}(2, \mathbb{C})$ and thus is an intertwining operator. By Schur lemma (Lemma 4.23), this shows that C acts by a constant in every irreducible representation. And if V is not irreducible, eigenspaces of V are subrepresentations, which could be used to decompose V into irreducible representations (see Lemma 4.21).

Element C is called the *Casimir operator* for $\mathfrak{sl}(2, \mathbb{C})$. We will discuss its generalization for other Lie algebras in Proposition 6.15.

Proposition 5.7.

(1) *The adjoint action of $\mathfrak{g}$ on $\mathfrak{g}$ can be uniquely extended to an action of $\mathfrak{g}$ on $U\mathfrak{g}$ which satisfies Leibniz rule:* $\operatorname{ad} x.(ab) = (\operatorname{ad} x.a)b + a(\operatorname{ad} x.b), x \in \mathfrak{g}, a, b \in U\mathfrak{g}$. *Moreover,* $\operatorname{ad} x.a = xa - ax$.
(2) *Let $Z\mathfrak{g} = Z(U\mathfrak{g})$ be the center of universal enveloping algebra. Then $Z\mathfrak{g}$ coincides with the space of invariants of $U\mathfrak{g}$ with respect to the adjoint action of $\mathfrak{g}$:*

$$Z\mathfrak{g} = (U\mathfrak{g})^{\operatorname{ad} \mathfrak{g}}.$$

Proof. Define adjoint action of $x \in \mathfrak{g}$ on $U\mathfrak{g}$ by $\operatorname{ad} x.a = xa - ax$. Clearly, for $a \in \mathfrak{g}$ this coincides with the usual definition of adjoint action. To see that it is indeed an action, we need to verify that $\operatorname{ad}[x, y].a =$

ad x(ad $y.a$) − ad y(ad $x.a$), or

$$[x, y]a - a[x, y] = \left(x(ya - ay) - (ya - ay)x\right) - (y(xa - ax)$$
$$-(xa - ax)y),$$

which is given by explicit calculation.

Leibniz rule follows from

$$xab - abx = (xa - ax)b + a(xb - bx).$$

This proves the first part. The second part follows immediately: $C \in U\mathfrak{g}$ is central iff it commutes with all the generators, i.e. if $Cx = xC$ for any $x \in \mathfrak{g}$. The last condition is equivalent to ad $x.C = 0$. □

5.2. Poincare–Birkhoff–Witt theorem

In this section, $\mathfrak{g}$ is a finite-dimensional Lie algebra over the field $\mathbb{K}$ and $U\mathfrak{g}$ is the universal enveloping algebra of $\mathfrak{g}$.

We had already mentioned that $U\mathfrak{g}$ is infinite-dimensional. In this section, we will give a more precise statement.

Unlike polynomial algebra, $U\mathfrak{g}$ is not graded: if we try to define degree by $\deg(x_1 \ldots x_k) = k, x_i \in \mathfrak{g}$, then we run into problem with the defining relation (5.1): we would have $\deg(xy) = \deg(yx) = 2$, but $\deg(xy - yx) = \deg([x, y]) = 1$. Instead, we have a weaker structure: we can define filtration on $U\mathfrak{g}$ by letting, for any $k \geq 0$,

$$U_k\mathfrak{g} = \text{Subspace in } U\mathfrak{g} \text{ spanned by products } x_1 \ldots x_p, \quad p \leq k. \qquad (5.4)$$

This defines a filtration on $U\mathfrak{g}$: we have

$$\mathbb{K} = U_0\mathfrak{g} \subset U_1\mathfrak{g} \subset \ldots, \qquad U\mathfrak{g} = \bigcup U_p\mathfrak{g}.$$

The following proposition gives some properties of this filtration.

Proposition 5.8.

(1) $U\mathfrak{g}$ *is a filtered algebra: if* $x \in U_p\mathfrak{g}, y \in U_q\mathfrak{g}$, *then* $xy \in U_{p+q}\mathfrak{g}$.
(2) *If* $x \in U_p\mathfrak{g}, y \in U_q\mathfrak{g}$, *then* $xy - yx \in U_{p+q-1}\mathfrak{g}$.
(3) *Let* $x_1, \ldots, x_n$ *be an ordered basis in* $\mathfrak{g}$. *Then monomials*

$$x_1^{k_1} \ldots x_n^{k_n}, \qquad \sum k_i \leq p \qquad (5.5)$$

span $U_p\mathfrak{g}$. *Note that we fix the order of basis elements.*

Proof. Part (1) is obvious. To prove the second part, note that for $p = 1$, we have

$$x(y_1 \ldots y_q) - (y_1 \ldots y_q)x = \sum_i y_1 \ldots [x, y_i] \ldots y_q \in U_q\mathfrak{g}.$$

In particular, this implies that for any $x \in \mathfrak{g}, y \in U_q\mathfrak{g}$, we have $xy \equiv yx$ mod $U_q\mathfrak{g}$.

Now we can argue by induction in p: if the statement is true for some p, then

$$x_1 \ldots x_p + 1y \equiv x_1 \ldots x_p y x_p + 1 \equiv y x_1 \ldots x_p x_p + 1 \quad \text{mod } U_{p+q}\mathfrak{g}.$$

Part (3) is again proved by induction in p. Indeed, for $p = 1$ it is obvious. To establish the induction step, notice that $U_{p+1}\mathfrak{g}$ is generated by elements of the form $xy, x \in \mathfrak{g}, y \in U_p\mathfrak{g}$. By induction assumption, y can be written as linear combination of monomials of the form (5.5). But by part (2),

$$x_i(x_1^{k_1} \ldots x_n^{k_n}) - x_1^{k_1} \ldots x_i^{k_i+1} \ldots x_n^{k_n} \in U_p\mathfrak{g}.$$

Using the induction assumption again, we see that $x_i(x_1^{k_1} \ldots x_n^{k_n})$ can again be written as linear combination of monomials of the form (5.5), with $\sum k_i \le p + 1$. $\qquad\square$

Corollary 5.9. *Each $U_p\mathfrak{g}$ is finite-dimensional.*

Corollary 5.10. *The associated graded algebra*

$$\text{Gr } U\mathfrak{g} = \bigoplus_p U_p\mathfrak{g}/U_{p-1}\mathfrak{g} \tag{5.6}$$

is commutative.

We can now formulate the main result of this section.

Theorem 5.11 (Poincaré–Birkhoff–Witt)**.** *Let $x_1, \ldots, x_n$ be an ordered basis in $\mathfrak{g}$. Then monomials of the form (5.5) form a basis in $U_p\mathfrak{g}$.*

The proof of this theorem is not given here; it can be found, for example, in [9, 22, 24]. Here is the main idea of the proof. Since we already know that monomials of the form (5.5) generate $U_p\mathfrak{g}$ (see Proposition 5.8), it suffices to show that they are linearly independent. To show this, we construct a representation in which the operators corresponding to these monomials are linearly independent.

Namely, we consider (infinite-dimensional) vector space V with basis given by (5.5) (no restriction on $\sum k_i$). The action is uniquely defined by the requirement that $\rho(x_i).x_{j_1} \ldots x_{j_n} = x_i x_{j_1} \ldots x_{j_n}$ if $i \le j_1 \le j_2 \ldots$. For example, this forces $\rho(x_1).x_2 = x_1 x_2$.

This requirement also determines $\rho(x_i).x_{j_1} \ldots x_{j_n}$ if $i > j_1$. For example, to define $\rho(x_2).x_1$, we note that it must be equal to $\rho(x_2)\rho(x_1).1 = \rho(x_1)\rho(x_2).1 + \rho([x_2, x_1]).1 = x_1 x_2 + \sum a_i x_i$, where a_i are defined by $[x_1, x_2] = \sum a_i x_i$.

The difficult part is to check that it is indeed an action, i.e., that it satisfies $\rho(x)\rho(y) - \rho(y)\rho(x) = \rho[x, y]$, which is done by an explicit calculation using the Jacobi identity.

Note that this theorem would fail without the Jacobi identity: if $[\,,\,]: \mathfrak{g} \otimes \mathfrak{g} \to \mathfrak{g}$ is an antisymmetric map not satisfying Jacobi identity, then the algebra defined by (5.1) can be trivial (i.e., all $i(x) = 0$).

This theorem can also be reformulated in a coordinate-independent way.

Theorem 5.12 (Poincaré–Birkhoff–Witt). *The graded algebra* $\operatorname{Gr} U\mathfrak{g}$ *defined by* (5.6) *is naturally isomorphic to the symmetric algebra* $S\mathfrak{g}$. *The isomorphism is given by*

$$S^p \mathfrak{g} \to \operatorname{Gr}^p U\mathfrak{g}$$
$$a_1 \ldots a_p \mapsto a_1 \ldots a_p \quad \mod U_{p-1}\mathfrak{g} \tag{5.7}$$

and the inverse isomorphism is given by

$$\operatorname{Gr}^p U\mathfrak{g} \to S^p \mathfrak{g}$$
$$a_1 \ldots a_p \mapsto a_1 \ldots a_p, \tag{5.8}$$
$$a_1 \ldots a_l \mapsto 0, \quad l < p.$$

When written in this form, this theorem may seem trivial. The non-triviality is hidden in the statement that the maps (5.7), (5.8) are well-defined.

The Poincaré–Birkhoff–Witt (or PBW for short) theorem has a number of useful corollaries. Here are some of them; proofs are left as an easy exercise to the reader.

Corollary 5.13. *The natural map* $\mathfrak{g} \to U\mathfrak{g}$ *is injective.*

Corollary 5.14. *Let* $\mathfrak{g}_1, \mathfrak{g}_2 \subset \mathfrak{g}$ *be subalgebras such that* $\mathfrak{g} = \mathfrak{g}_1 \oplus \mathfrak{g}_2$ *as a vector space (we do not require that* $\mathfrak{g}_1, \mathfrak{g}_2$ *commute). Then the multiplication map*

$$U\mathfrak{g}_1 \otimes U\mathfrak{g}_2 \to U\mathfrak{g}$$

is a vector space isomorphism.

Corollary 5.15. *Algebra* $U\mathfrak{g}$ *has no zero divisors.*

Notice that while Theorem 5.12 establishes an isomorphism between $\text{Gr}\,U\mathfrak{g}$ and $S\mathfrak{g}$, this isomorphism clearly can not be extended to an isomorphism of algebras $U\mathfrak{g} \xrightarrow{\sim} S\mathfrak{g}$ unless $\mathfrak{g}$ is commutative. The following result, the proof of which is left as an exercise for the reader, is the best one can get in this direction for general $\mathfrak{g}$.

Theorem 5.16. *The map* $S\mathfrak{g} \to U\mathfrak{g}$ *given by*

$$\mathbf{sym}(x_1 \ldots x_p) = \frac{1}{p!} \sum_{s \in S_p} x_{s(1)} \ldots x_{s(p)} \tag{5.9}$$

is an isomorphism of $\mathfrak{g}$*-modules.*

This isomorphism will be later used in the construction of so-called Harish–Chandra isomorphism (see Section 8.8).

5.3. Ideals and commutant

Recall that a subalgebra of $\mathfrak{g}$ is a vector subspace closed under the commutator, and an ideal is a vector subspace $\mathfrak{h}$ such that $[x, y] \in \mathfrak{h}$ for any $x \in \mathfrak{g}, y \in \mathfrak{h}$. This definition is the natural analog of an ideal in an associative algebra. Note, however, that because of skew-symmetry of the commutator there is no difference between left and right ideals: every right ideal is also automatically a left ideal.

As in the theory of associative algebras, if $\mathfrak{h}$ is an ideal of $\mathfrak{g}$ then the quotient space $\mathfrak{g}/\mathfrak{h}$ has a canonical structure of a Lie algebra, and we have the following trivial result, proof of which is left to the reader as an exercise.

Lemma 5.17. *If* $f : \mathfrak{g}_1 \to \mathfrak{g}_2$ *is a morphism of Lie algebras, then* $\text{Ker}\,f$ *is an ideal in* $\mathfrak{g}_1$, $\text{Im}\,f$ *is a subalgebra in* $\mathfrak{g}_2$, *and* f *gives rise to an isomorphism of Lie algebras* $\mathfrak{g}_1/\text{Ker}\,f \simeq \text{Im}\,f$.

In addition, here is another important result about ideals, proof of which is left as an easy exercise to the reader.

Lemma 5.18. *Let* I_1, I_2 *be ideals in* $\mathfrak{g}$. *Define*

$$I_1 + I_2 = \{x + y \mid x \in I_1, y \in I_2\}$$
$$[I_1, I_2] = \text{Subspace spanned by } [x, y], \quad x \in I_1, y \in I_2.$$

Then $I_1 \cap I_2, I_1 + I_2, [I_1, I_2]$ *are ideals in* $\mathfrak{g}$.

One of the first ways to study Lie algebras is by analyzing how close the Lie algebra is to a commutative Lie algebra. There are several ways of making it precise.

First, we might look at how large the center $\mathfrak{z}(\mathfrak{g}) = \{x \in \mathfrak{g} \mid [x, y] = 0 \text{ for all } y \in \mathfrak{g}\}$ is. However, it turns out that it is more effective to study commutative quotients of $\mathfrak{g}$.

Definition 5.19. The commutant of a Lie algebra $\mathfrak{g}$ is the ideal $[\mathfrak{g}, \mathfrak{g}]$.

The following lemma explains the importance of the commutant.

Lemma 5.20. *The quotient $\mathfrak{g}/[\mathfrak{g}, \mathfrak{g}]$ is an abelian Lie algebra. Moreover, $[\mathfrak{g}, \mathfrak{g}]$ is the smallest ideal with this property: if $\mathfrak{g}/I$ is abelian for some ideal $I \subset \mathfrak{g}$, then $I \supset [\mathfrak{g}, \mathfrak{g}]$.*

Commutant gives us another way of measuring how far a Lie algebra is from being commutative: the smaller $[\mathfrak{g}, \mathfrak{g}]$ (and the larger $\mathfrak{g}/[\mathfrak{g}, \mathfrak{g}]$), the closer $\mathfrak{g}$ is to being commutative. For example, for commutative $\mathfrak{g}$, we have $[\mathfrak{g}, \mathfrak{g}] = 0$.

Example 5.21. The commutant $[\mathfrak{gl}(n, \mathbb{K}), \mathfrak{gl}(n, \mathbb{K})] = [\mathfrak{sl}(n, \mathbb{K}), \mathfrak{sl}(n, \mathbb{K})] = \mathfrak{sl}(n, \mathbb{K})$. Indeed, it is obvious that for any $z = [x, y]$ we have $\operatorname{tr} z = 0$. On the other hand, for $i \neq j$ we have $E_{ii} - E_{jj} = [E_{ij}, E_{ji}]$ and $2E_{ij} = [E_{ii} - E_{jj}, E_{ij}]$, which shows that $E_{ii} - E_{jj}, E_{ij} \in [\mathfrak{sl}(n, \mathbb{K}), \mathfrak{sl}(n, \mathbb{K})]$. Since these elements span $\mathfrak{sl}(n, \mathbb{K})$ we see that $[\mathfrak{gl}(n, \mathbb{K}), \mathfrak{gl}(n, \mathbb{K})] = [\mathfrak{sl}(n, \mathbb{K}), \mathfrak{sl}(n, \mathbb{K})] = \mathfrak{sl}(n, \mathbb{K})$.

5.4. Solvable and nilpotent Lie algebras

We now can define an important class of Lie algebras.

Definition 5.22. For a Lie algebra $\mathfrak{g}$, define the series of ideals $D^i\mathfrak{g}$ (called the *derived series*) by $D^0\mathfrak{g} = \mathfrak{g}$ and

$$D^{i+1}\mathfrak{g} = [D^i\mathfrak{g}, D^i\mathfrak{g}].$$

It immediately follows from Lemmas 5.18 and 5.20 that each D^i is an ideal in $\mathfrak{g}$ and $D^i\mathfrak{g}/D^{i+1}\mathfrak{g}$ is abelian.

Proposition 5.23. *The following conditions are equivalent:*

(1) $D^n\mathfrak{g} = 0$ *for large enough n.*
(2) *There exists a sequence of subalgebras $\mathfrak{a}^0 = \mathfrak{g} \supset \mathfrak{a}^1 \supset \cdots \supset \mathfrak{a}^k = \{0\}$ such that $\mathfrak{a}^{i+1}$ is an ideal in $\mathfrak{a}^i$ and the quotient $\mathfrak{a}^i/\mathfrak{a}^{i+1}$ is abelian.*

(3) *For large enough n, every commutator of the form*

$$[\ldots [[x_1, x_2], [x_3, x_4]] \ldots]$$

(2^n terms, arranged in a binary tree of length n) is zero.

Proof. Equivalence of (1) and (3) is obvious. Implication (1) $\Longrightarrow$ (2) is also clear: we can take $\mathfrak{a}^i = D^i\mathfrak{g}$. To prove (2) $\Longrightarrow$ (1), note that if $\mathfrak{a}^i$ satisfies the conditions of the proposition, then by Lemma 5.20, we have $\mathfrak{a}^{i+1} \supset [\mathfrak{a}^i, \mathfrak{a}^i]$. Thus, reasoning by induction, we see that $\mathfrak{a}^i \supset D^i\mathfrak{g}$.   □

Definition 5.24. Lie algebra $\mathfrak{g}$ is called *solvable* if it satisfies any of the equivalent conditions of Proposition 5.23.

Informally, a solvable Lie algebra is an "almost commutative" Lie algebra: it is an algebra that can be obtained by successive extensions of commutative algebras.

This is not the only way of making the notion of "almost commutative" Lie algebra precise. Another class of Lie algebras can be defined as follows.

Definition 5.25. For a Lie algebra $\mathfrak{g}$, define a series of ideals $D_i\mathfrak{g} \subset \mathfrak{g}$ (called *lower central series*) by $D_0\mathfrak{g} = \mathfrak{g}$ and

$$D_{i+1}\mathfrak{g} = [\mathfrak{g}, D_i\mathfrak{g}].$$

Proposition 5.26. *The following conditions are equivalent:*

(1) *$D_n\mathfrak{g} = 0$ for large enough n.*
(2) *There exists a sequence of ideals $\mathfrak{a}_0 = \mathfrak{g} \supset \mathfrak{a}_1 \supset \cdots \supset \mathfrak{a}_k = \{0\}$ such that $[\mathfrak{g}, \mathfrak{a}_i] \subset \mathfrak{a}_{i+1}$.*
(3) *For large enough n, every commutator of the form*

$$[\ldots [[x_1, x_2], x_3], x_4] \ldots x_n]$$

(n terms) is zero.

Proof. Equivalence of (1) and (3) is obvious. Implication (1) $\Longrightarrow$ (2) is also clear: we can take $\mathfrak{a}_i = D_i\mathfrak{g}$. To prove (2) $\Longrightarrow$ (1), note that if $\mathfrak{a}_i$ satisfies the conditions of the proposition, then by induction, we see that $\mathfrak{a}_i \supset D_i\mathfrak{g}$.   □

Definition 5.27. Lie algebra $\mathfrak{g}$ is called *nilpotent* if it satisfies any of the equivalent conditions of Proposition 5.26.

Example 5.28. Let $\mathfrak{b} \subset \mathfrak{gl}(n, \mathbb{K})$ be the subalgebra of upper triangular matrices, and $\mathfrak{n}$ be the subalgebra of all strictly upper triangular matrices. Then $\mathfrak{b}$ is solvable, and $\mathfrak{n}$ is nilpotent.

To prove it, let us first generalize it. Namely, if $\mathcal{F}$ is a flag in a finite-dimensional vector space V:

$$\mathcal{F} = (\{0\} \subset V_1 \subset V_2 \subset \ldots V_n = V)$$

with $\dim V_i < \dim V_{i+1}$ (we do not require that $\dim V_i = i$), then define

$$\mathfrak{b}(\mathcal{F}) = \{x \in \mathfrak{gl}(V) \mid xV_i \subset V_i \text{ for all } i\},$$
$$\mathfrak{n}(\mathcal{F}) = \{x \in \mathfrak{gl}(V) \mid xV_i \subset V_{i-1} \text{ for all } i\}.$$

By taking $\mathcal{F}$ to be the standard flag in $\mathbb{K}^n$ (see Example 2.25) we recover the Lie algebras $\mathfrak{b}, \mathfrak{n}$ defined above.

We claim that $\mathfrak{n}(\mathcal{F})$ is nilpotent. Indeed, define more general algebras

$$\mathfrak{a}_k(\mathcal{F}) = \{x \in \mathfrak{gl}(V) \mid xV_i \subset V_{i-k} \text{ for all } i\}$$

so that $\mathfrak{b}(\mathcal{F}) = \mathfrak{a}_0, \mathfrak{n}(\mathcal{F}) = \mathfrak{a}_1$. Then it is obvious that for $x \in \mathfrak{a}_k, y \in \mathfrak{a}_l$, we have $xy \in \mathfrak{a}_{k+l}$ (here xy is the usual product in $\operatorname{End}(V)$); thus, $[\mathfrak{a}_k, \mathfrak{a}_l] \subset \mathfrak{a}_{k+l}$, so $D_i\mathfrak{n} \subset \mathfrak{a}_{i+1}$. This proves nilpotency of $\mathfrak{n}(\mathcal{F})$.

To show solvability of $\mathfrak{b}$ (for the standard flag $\mathcal{F}$), note that even though for $x, y \in \mathfrak{b}$ we can only say that $xy \in \mathfrak{b}$, for the commutator we have a stronger condition: $[x, y] \in \mathfrak{n} = \mathfrak{a}_1$. Indeed, diagonal entries of xy and yx coincide. Thus, $D^1\mathfrak{b} \subset \mathfrak{n} = \mathfrak{a}_1$. From here it easily follows by induction that $D^{i+1}\mathfrak{b} \subset \mathfrak{a}_{2^i}$.

Note, finally, that $\mathfrak{b}$ is not nilpotent: $D_2\mathfrak{b} = [\mathfrak{b}, D_1\mathfrak{b}] = D_1\mathfrak{b} = \mathfrak{n}$, which can be easily deduced from $[x, E_{ij}] = (\lambda_i - \lambda_j)E_{ij}$ if x is a diagonal matrix with entries λ_i.

The following theorem summarizes some basic properties of solvable and nilpotent Lie algebras.

Theorem 5.29.

(1) *A real Lie algebra $\mathfrak{g}$ is solvable (respectively, nilpotent) iff its complexification $\mathfrak{g}_{\mathbb{C}}$ is solvable (respectively, nilpotent).*
(2) *If $\mathfrak{g}$ is solvable, then any subalgebra and quotient of $\mathfrak{g}$ are also solvable. If $\mathfrak{g}$ is nilpotent, then any subalgebra, quotient of $\mathfrak{g}$ is also nilpotent.*
(3) *If $\mathfrak{g}$ is nilpotent, then $\mathfrak{g}$ is solvable.*
(4) *If $I \subset \mathfrak{g}$ is an ideal such that both $I, \mathfrak{g}/I$ are solvable, then $\mathfrak{g}$ is solvable.*

Proof. Parts (1), (2) are obvious if we use definition of solvable algebra in the form "any commutator of the form ... is zero", and similarly for nilpotent. Part (3) follows from inclusion $D^i \mathfrak{g} \subset D_i \mathfrak{g}$, which can be easily proved by induction.

Finally, to prove part (4), denote by φ the canonical projection $\mathfrak{g} \to \mathfrak{g}/I$. Then $\varphi(D^n \mathfrak{g}) = D^n(\mathfrak{g}/I) = 0$ for some n. Thus, $D^n \mathfrak{g} \subset I$. Therefore, $D^{n+k} \mathfrak{g} \subset D^k I$, so $D^{n+k} \mathfrak{g} = 0$ for large enough k. $\square$

5.5. Lie's and Engel's theorems

The main result of this section is the following theorem.

Theorem 5.30 (Lie's theorem about representations of a solvable Lie algebra). *Let $\rho \colon \mathfrak{g} \to \mathfrak{gl}(V)$ be a complex representation of a solvable Lie algebra $\mathfrak{g}$ (real or complex). Then there exists a basis in V such that in this basis, all operators $\rho(x)$ are upper-triangular.*

This theorem is a generalization of a well-known result that any operator in a complex vector space can be brought to an upper-triangular form by a change of basis.

The key step in the proof of the theorem is the following result.

Proposition 5.31. *Let $\rho \colon \mathfrak{g} \to \mathfrak{gl}(V)$ be a complex representation of a solvable Lie algebra $\mathfrak{g}$. Then there exists a vector $v \in V$ which is a common eigenvector of all $\rho(x), x \in \mathfrak{g}$.*

Proof. The proof goes by induction in dimension of $\mathfrak{g}$. Since $\mathfrak{g}$ is solvable, $[\mathfrak{g}, \mathfrak{g}] \neq \mathfrak{g}$. Let $\mathfrak{g}' \subset \mathfrak{g}$ be a subspace which contains $[\mathfrak{g}, \mathfrak{g}]$ and has codimension 1 in $\mathfrak{g} \colon \mathfrak{g} = \mathfrak{g}' \oplus \mathbb{C}x$. Then $\mathfrak{g}'$ is an ideal in $\mathfrak{g}$; thus, $\mathfrak{g}'$ is solvable.

By induction assumption, there exists $v \in V$ which is a common eigenvector for all $\rho(h), h \in \mathfrak{g}' \colon \rho(h)v = \lambda(h)v$. Consider the vector space W spanned by $v^0 = v, v^1 = \rho(x)v, v^2 = (\rho(x))^2 v, \ldots$.

We claim that W is stable under the action of any $h \in \mathfrak{g}'$; moreover,

$$hv^k = \lambda(h)v^k + \sum_{l<k} a_{kl}(h)v^l. \tag{5.10}$$

This is easily proved by induction: indeed,

$$hv^k = hxv^{k-1} = xhv^{k-1} + [h,x]v^{k-1} = \lambda(h)xv^{k-1}$$
$$+ \lambda([h,x])v^{k-1} + \cdots \tag{5.11}$$

Thus, W is stable under the action of $\mathfrak{g}$.

Let n be the smallest integer such that v^{n+1} is in the subspace generated by $v^0, v^1, \ldots, v^n$. Then $v^0, v^1, \ldots, v^n$ is a basis in W. By (5.10), in this basis any $\rho(h)$ is upper-triangular, with $\lambda(h)$ on the diagonal. In particular, this implies that $\operatorname{tr}_W \rho(h) = (n+1)\lambda(h)$.

Since $\operatorname{tr}_W[\rho(x), \rho(h)] = 0$, this implies that $\lambda([h, x]) = 0$ for any $h \in \mathfrak{g}'$. The same calculation as in (5.11), shows that this implies $hv^k = \lambda(h)v^k$. Therefore, any vector $w \in W$ is a common eigenvector for all $h \in \mathfrak{g}'$. Choosing w to be an eigenvector for x, we get the statement of the proposition. $\square$

This proposition immediately implies Lie's theorem.

Proof of Theorem 5.30. Proof goes by induction in $\dim V$. By Proposition 5.31, there exists a common eigenvector v for all $x \in \mathfrak{g}$. Consider the space $V/\mathbb{C}v$. By induction assumption, there exists a basis $v_1, v_2, \ldots$ in $V/\mathbb{C}v$ such that the action of $\mathfrak{g}$ in this basis of $V/\mathbb{C}v$ is upper-triangular. For each of these vectors, choose a preimage $\tilde{v}_i \in V$. Then one immediately sees that the action of any $x \in \mathfrak{g}$ in the basis $v, \tilde{v}_1, \tilde{v}_2, \ldots$ is upper-triangular. $\square$

This theorem gives a number of useful corollaries.

Corollary 5.32.

(1) *Any irreducible complex representation of a solvable Lie algebra is one-dimensional.*

(2) *If a complex Lie algebra $\mathfrak{g}$ is solvable, then there exists a sequence $0 \subset I_1 \subset \cdots \subset I_n = \mathfrak{g}$, where each I_k is an ideal in $\mathfrak{g}$ and I_{k+1}/I_k is one-dimensional.*

(3) *$\mathfrak{g}$ is solvable if and only if $[\mathfrak{g}, \mathfrak{g}]$ is nilpotent.*

Proof. Part (1) is obvious from Proposition 5.31; part (2) is immediately obtained if we apply Lie's theorem to the adjoint representation and note that a subrepresentation of the adjoint representation is the same as an ideal in $\mathfrak{g}$.

To prove part (3), note that implication in one direction is obvious. Indeed, if $[\mathfrak{g}, \mathfrak{g}]$ is nilpotent, then it is also solvable; since $\mathfrak{g}/[\mathfrak{g}, \mathfrak{g}]$ is commutative (and thus solvable), by Theorem 5.29, $\mathfrak{g}$ itself is solvable.

Conversely, assume that $\mathfrak{g}$ is solvable. Without loss of generality, we may assume that $\mathfrak{g}$ is complex. Apply Lie's theorem to the adjoint representation. By Theorem 5.30, $\operatorname{ad} \mathfrak{g} \subset \mathfrak{b}$ (algebra of upper-triangular matrices) in some basis of $\mathfrak{g}$; thus, by results of Example 5.28, the algebra $[\operatorname{ad}\mathfrak{g}, \operatorname{ad}\mathfrak{g}] = \operatorname{ad}[\mathfrak{g}, \mathfrak{g}] \subset \mathfrak{n}$ is nilpotent, so $\operatorname{ad}[x_1, [\ldots [x_{n-1}, x_n] \ldots] = 0$ for sufficiently large n and all $x_i \in [\mathfrak{g}, \mathfrak{g}]$. Thus, $[y, [x_1, [\ldots [x_{n-1}, x_n] \ldots] = 0$ for sufficiently large n and all $x_i, y \in [\mathfrak{g}, \mathfrak{g}]$. $\square$

One also might ask if there is an analog of Lie's theorem for nilpotent Lie algebras. Of course, since every nilpotent Lie algebra is automatically solvable (Theorem 5.29), Lie's theorem shows that in any representation of a nilpotent algebra, operators $\rho(x)$ are upper-triangular in a certain basis. One wonders whether one has a stronger result – for example, whether operators $\rho(x)$ can be made strictly upper-triangular. Here the answer is obviously negative: it suffices to take a commutative Lie algebra which acts diagonally in $\mathbb{C}^n$.

The proper analog of Lie's theorem for nilpotent Lie algebras is given by the following result.

Theorem 5.33. *Let V be a finite-dimensional vector space, either real or complex, and let $\mathfrak{g} \subset \mathfrak{gl}(V)$ be a Lie subalgebra which consists of nilpotent operators. Then there exists a basis in V such that all operators $x \in \mathfrak{g}$ are strictly upper-triangular.*

The proof of this theorem will not be given here; interested reader can find it in [46], [24], or [22]. It is not very difficult and in fact is rather similar to the proof of Lie's theorem; the only reason it is not given here is because it does not give any new insight.

As an immediate corollary, we get the following theorem.

Theorem 5.34 (Engel's theorem). *A Lie algebra $\mathfrak{g}$ is nilpotent if and only if for every $x \in \mathfrak{g}$, the operator $\operatorname{ad} x \in \operatorname{End}(\mathfrak{g})$ is nilpotent.*

Proof. One direction is obvious: if $\mathfrak{g}$ is nilpotent then by definition, $[x, [x, \ldots [x, y \ldots] = (\operatorname{ad} x)^n . y = 0$ for large enough n.

Conversely, if $\operatorname{ad} x$ is nilpotent for every x, then by the previous theorem, there exists a sequence of subspaces $0 \subset \mathfrak{g}_1 \subset \mathfrak{g}_2 \cdots \subset \mathfrak{g}_n = \mathfrak{g}$ such that $\operatorname{ad} x.\mathfrak{g}_i \subset \mathfrak{g}_{i-1}$. This shows that each $\mathfrak{g}_i$ is an ideal in $\mathfrak{g}$ and moreover, $[\mathfrak{g}, \mathfrak{g}_i] \subset \mathfrak{g}_{i-1}$. Thus, $\mathfrak{g}$ is nilpotent. $\square$

5.6. The radical. Semisimple and reductive algebras

So far, we have defined the notion of a solvable Lie algebra; informally, a solvable Lie algebra is the one which is close to being abelian. In this section, we will describe the opposite extreme case, Lie algebras which are as far as possible from being abelian (they are called semisimple) and show that in a reasonable sense, any Lie algebra is built out of a solvable and semisimple one.

Definition 5.35. A Lie algebra $\mathfrak{g}$ is called semisimple if it contains no nonzero solvable ideals.

Note that this in particular implies that the center $\mathfrak{z}(\mathfrak{g}) = 0$.

A special case of semisimple Lie algebras is given by simple ones.

Definition 5.36. A Lie algebra $\mathfrak{g}$ is called simple if it is not abelian and contains no ideals other than 0 and $\mathfrak{g}$.

The condition that $\mathfrak{g}$ should not be abelian is included to rule out the one-dimensional Lie algebra: there are many reasons not to include it in the class of simple Lie algebras. One of these reasons is the following lemma.

Lemma 5.37. *Any simple Lie algebra is semisimple.*

Proof. If $\mathfrak{g}$ is simple, then it contains no ideals other than 0 and $\mathfrak{g}$. Thus, if $\mathfrak{g}$ contains a nonzero solvable ideal, then it must coincide with $\mathfrak{g}$, so $\mathfrak{g}$ must be solvable. But then $[\mathfrak{g}, \mathfrak{g}]$ is an ideal which is strictly smaller than $\mathfrak{g}$ (because $\mathfrak{g}$ is solvable) and nonzero (because $\mathfrak{g}$ is not abelian). This gives a contradiction. $\square$

Example 5.38. The Lie algebra $\mathfrak{sl}(2, \mathbb{C})$ is simple. Indeed, recall that $\operatorname{ad} h$ is diagonal in the basis e, f, h, with eigenvalues $2, -2, 0$ (see Section 3.10). Any ideal in $\mathfrak{g}$ must be stable under $\operatorname{ad} h$. Now we can use the following easy to prove result from linear algebra: if A is a diagonalizable operator in a finite-dimensional vector space, with distinct eigenvalues: $Av_i = \lambda_i v_i, \lambda_i \neq \lambda_j$, then the only subspaces invariant under A are those spanned by some of the eigenvectors v_i. Applying this to $\operatorname{ad} h$, we see that any ideal in $\mathfrak{sl}(2, \mathbb{C})$ must be spanned as a vector space by a subset of $\{e, f, h\}$.

But if an ideal I contains h, then $[h, e] = 2e \in I, [h, f] = -2f \in I$, so $I = \mathfrak{sl}(2, \mathbb{C})$. If I contains e, then $[e, f] = h \in I$, so again $I = \mathfrak{sl}(2, \mathbb{C})$. Similarly, if $f \in I$, then $I = \mathfrak{sl}(2, \mathbb{C})$. Thus, $\mathfrak{sl}(2, \mathbb{C})$ contains no non-trivial ideals.

In the next section, we will generalize this result and show that classical Lie algebras such as $\mathfrak{sl}(n, \mathbb{C}), \mathfrak{su}(n), \mathfrak{sp}(n, \mathbb{C}), \mathfrak{so}(n, \mathbb{C})$ are semisimple.

For a general Lie algebra $\mathfrak{g}$, which is neither semisimple nor solvable, we can try to "separate" the solvable and semisimple parts.

Proposition 5.39. *In any Lie algebra $\mathfrak{g}$, there is a unique solvable ideal which contains any other solvable ideal. This solvable ideal is called the* radical *of $\mathfrak{g}$ and denoted by* $\operatorname{rad}(\mathfrak{g})$.

Proof. Uniqueness is obvious. To show existence, note that if I_1, I_2 are solvable ideals, then so is $I_1 + I_2$. Indeed, it contains solvable ideal I_1 and the quotient $(I_1 + I_2)/I_1 = I_2/(I_1 \cap I_2)$ is also solvable since it is a quotient of I_2. Thus, by Theorem 5.29, $I_1 + I_2$ is also solvable. By induction, this shows that any finite sum of solvable ideals is also solvable. Thus, we can let $\operatorname{rad}(\mathfrak{g}) = \sum I$, where the sum is taken over all solvable ideals (finite-dimensionality of $\mathfrak{g}$ shows that it suffices to take a finite sum). $\qquad\qquad\Box$

Using this definition, we can rewrite the definition of a semisimple Lie algebra as follows: $\mathfrak{g}$ is semisimple iff $\operatorname{rad}(\mathfrak{g}) = 0$.

Theorem 5.40. *For any Lie algebra $\mathfrak{g}$, the quotient $\mathfrak{g}/\operatorname{rad}(\mathfrak{g})$ is semisimple. Conversely, if $\mathfrak{b}$ is a solvable ideal in $\mathfrak{g}$ such that $\mathfrak{g}/\mathfrak{b}$ is semisimple, then $\mathfrak{b} = \operatorname{rad}(\mathfrak{g})$.*

Proof. Assume that $\mathfrak{g}/\operatorname{rad}(\mathfrak{g})$ contains a solvable ideal I. Consider the ideal $\tilde{I} = \pi^{-1}(I) \subset \mathfrak{g}$, where π is the canonical map $\mathfrak{g} \to \mathfrak{g}/\operatorname{rad}(\mathfrak{g})$. Then $\tilde{I} \supset \operatorname{rad}(\mathfrak{g})$ and $\tilde{I}/\operatorname{rad}(\mathfrak{g}) = I$ is solvable. Thus, by Theorem 5.29, $\tilde{I}$ is solvable, so $\tilde{I} = \operatorname{rad}(\mathfrak{g}), I = 0$.

Proof of the second statement is left to the reader as an exercise. $\qquad\Box$

This theorem shows that any Lie algebra can be included in a short exact sequence $0 \to \mathfrak{b} \to \mathfrak{g} \to \mathfrak{g}_{ss} \to 0$, where $\mathfrak{b}$ is solvable and $\mathfrak{g}_{ss}$ is semisimple. In fact, one has a much stronger result.

Theorem 5.41 (Levi theorem). *Any Lie algebra can be written as a direct sum*

$$\mathfrak{g} = \operatorname{rad}(\mathfrak{g}) \oplus \mathfrak{g}_{ss}, \tag{5.12}$$

where $\mathfrak{g}_{ss}$ is a semisimple subalgebra (not necessarily an ideal!) in $\mathfrak{g}$. Such a decomposition is called the Levi *decomposition for $\mathfrak{g}$.*

This theorem will not be proved here. A proof can be found in standard textbooks on Lie algebras, such as [46] or [24]. We only mention here that the key step in the proof is showing vanishing of a certain cohomology group; we will say more about this in Section 6.3.

Example 5.42. Let $G = \operatorname{SO}(3, \mathbb{R}) \ltimes \mathbb{R}^3$ be the Poincare group, i.e. the group of all maps $\mathbb{R}^3 \to \mathbb{R}^3$ which have the form $x \mapsto Ax + b, A \in \operatorname{SO}(3, \mathbb{R}), b \in \mathbb{R}^3$. The corresponding Lie algebra is $\mathfrak{g} = \mathfrak{so}(3, \mathbb{R}) \oplus \mathbb{R}^3$, where the commutator is given by $[(A_1, b_1), (A_2, b_2)] = ([A_1, A_2], A_1 b_2 - A_2 b_1)$. Thus, $\mathbb{R}^3$ is an ideal and $\mathfrak{so}(3, \mathbb{R})$ is a subalgebra. Since $\mathbb{R}^3$ is abelian and $\mathfrak{so}(3, \mathbb{R})$ is semisimple (which follows from semisimplicity of $\mathfrak{so}(3, \mathbb{R})_{\mathbb{C}} \simeq \mathfrak{sl}(2, \mathbb{C})$, see Example 5.38), we see that $\mathfrak{g} = \mathfrak{so}(3, \mathbb{R}) \oplus \mathbb{R}^3$ is exactly the Levi decomposition.

Another instructive example of the Levi decomposition is the Levi decomposition for parabolic subalgebras; a special case is given in Exercise 5.3.

As in the theory of associative algebras, there is a relation between the radical of $\mathfrak{g}$ and kernels of irreducible representations.

Theorem 5.43. *Let V be an irreducible complex representation of* $\mathfrak{g}$. *Then any* $h \in \text{rad}(\mathfrak{g})$ *acts in V by scalar operators:* $\rho(h) = \lambda(h)\,\text{id}$. *Also, any* $h \in [\mathfrak{g}, \text{rad}(\mathfrak{g})]$ *acts by zero.*

Proof. By Proposition 5.31, there is a common eigenvector in V for all $h \in \text{rad}(\mathfrak{g})$:$\rho(h).v = \lambda(h)v$ for some λ: $\text{rad}(\mathfrak{g}) \to \mathbb{C}$. Define $V_\lambda = \{w \in V \mid \rho(h)w = \lambda(h)w$ for all $h \in \text{rad}(\mathfrak{g})\}$. Then the same argument as in the proof of Proposition 5.31 shows that for any $x \in \mathfrak{g}$, one has $\rho(x)(V_\lambda) \subset V_\lambda$. Thus, V_λ is a subrepresentation; since it is non-zero and V is irreducible, we must have $V = V_\lambda$, which proves the first statement of the theorem. The second statement immediately follows from the first one. $\square$

From the point of view of representation theory, having non-zero elements which act by zero in any irreducible representation significantly complicates the theory. Thus, it is natural to consider a class of algebras for which $[\mathfrak{g}, \text{rad}(\mathfrak{g})] = 0$.

Definition 5.44. A Lie algebra is called *reductive* if $\text{rad}(\mathfrak{g}) = \mathfrak{z}(\mathfrak{g})$, i.e. if $\mathfrak{g}/\mathfrak{z}(\mathfrak{g})$ is semisimple. (Recall that $\mathfrak{z}(\mathfrak{g})$ is the center of $\mathfrak{g}$.)

Of course, any semisimple Lie algebra is reductive (because then $\text{rad}(\mathfrak{g}) = \mathfrak{z}(\mathfrak{g}) = 0$), but the converse is not true: for example, any Lie algebra which is a direct sum of an abelian and semisimple algebras

$$\mathfrak{g} = \mathfrak{z} \oplus \mathfrak{g}_{ss}, \qquad [\mathfrak{z}, \mathfrak{g}_{ss}] = 0, \tag{5.13}$$

is reductive. In fact, it follows from the Levi theorem that any reductive Lie algebra must have such form. Later we will give an alternative proof of this result, which does not use the Levi theorem (see Theorem 6.24).

In the next section we will show that many classical Lie algebras such as $\mathfrak{gl}(n, \mathbb{C})$ or $\mathfrak{u}(n)$ are reductive.

5.7. Invariant bilinear forms and semisimplicity of classical Lie algebras

So far, we have only one example of a semisimple Lie algebra, namely $\mathfrak{sl}(2, \mathbb{C})$ (see Example 5.38), and the proof of its semisimplicity was done by brute force, by analyzing all possibilities for an ideal. It is clear that such a proof would be difficult to generalize to higher-dimensional Lie algebras: we need better tools.

The standard approach to the study of semisimplicity is based on the notion of invariant bilinear form. Recall that a bilinear form B on $\mathfrak{g}$ is called invariant if

$$B(\text{ad}\,x.y, z) + B(y, \text{ad}\,x.z) = 0$$

for any $x, y, z \in \mathfrak{g}$ (see Example 4.15). The following lemma shows the importance of such forms.

Lemma 5.45. *Let B be an invariant bilinear form on $\mathfrak{g}$, and $I \subset \mathfrak{g}$ an ideal. Let $I^{\perp}$ be the orthogonal complement of I with respect to B: $I^{\perp} = \{x \in \mathfrak{g} \mid B(x, y) = 0$ for all $y \in I\}$. Then $I^{\perp}$ is also an ideal in $\mathfrak{g}$. In particular, $\mathrm{Ker}\, B = \mathfrak{g}^{\perp}$ is an ideal in $\mathfrak{g}$.*

The proof of this lemma is trivial and left to the reader. Note, however, that in general we can not write $\mathfrak{g} = I \oplus I^{\perp}$, as it is quite possible that $I \cap I^{\perp} \neq 0$, even for a non-degenerate form B.

Example 5.46. Let $\mathfrak{g} = \mathfrak{gl}(n, \mathbb{C})$ and define the form by $B(x, y) = \mathrm{tr}(xy)$. Then it is a symmetric invariant bilinear form on $\mathfrak{g}$. Indeed, symmetry is well-known and invariance follows from the following identity

$$\mathrm{tr}([x, y]z + y[x, z]) = \mathrm{tr}(xyz - yxz + yxz - yzx) = \mathrm{tr}(xyz - yzx) = 0.$$

In fact, there is an even easier proof: since $\mathrm{tr}(gxg^{-1}gyg^{-1}) = \mathrm{tr}(gxyg^{-1}) = \mathrm{tr}(xy)$ for any $g \in \mathrm{GL}(n, \mathbb{C})$, we see that this form is invariant under the adjoint action of $\mathrm{GL}(n, \mathbb{C})$ which is equivalent to the invariance under the action of $\mathfrak{gl}(n, \mathbb{C})$.

This example can be easily generalized.

Proposition 5.47. *Let V be a representation of $\mathfrak{g}$ and define a bilinear form on $\mathfrak{g}$ by*

$$B_V(x, y) = \mathrm{tr}_V(\rho(x)\rho(y)). \tag{5.14}$$

Then B_V is a symmetric invariant bilinear form on $\mathfrak{g}$.

The proof is identical to the proof in Example 5.46.

However, this form can be degenerate or even zero. It turns out, however, that there is a close relation between non-degeneracy of such forms and semisimplicity of $\mathfrak{g}$.

Theorem 5.48. *Let $\mathfrak{g}$ be a Lie algebra with a representation V such that the form B_V defined by (5.14) is non-degenerate. Then $\mathfrak{g}$ is reductive.*

Proof. It suffices to show that $[\mathfrak{g}, \mathrm{rad}(\mathfrak{g})] = 0$. Let $x \in [\mathfrak{g}, \mathrm{rad}(\mathfrak{g})]$; then, by Theorem 5.43, x acts by zero in any irreducible representation V_i and thus $x \in \mathrm{Ker}\, B_{V_i}$. But if we have a short exact sequence of representations $0 \to V_1 \to W \to V_2 \to 0$, then $B_W = B_{V_1} + B_{V_2}$ (see Exercise 5.1). Thus, arguing by induction it is easy to see that for any representation V, we would have $x \in \operatorname{Ker} B_V$. Since by assumption B_V is non-degenerate, this shows $x = 0$. $\qquad\square$

As an immediate corollary, we have the following important result.

Theorem 5.49. *All classical Lie algebras of Section 2.7 are reductive. Algebras* $\mathfrak{sl}(n, \mathbb{K}), \mathfrak{so}(n, \mathbb{K})$ $(n > 2), \mathfrak{su}(n), \mathfrak{sp}(n, \mathbb{K})$ *are semisimple; algebras* $\mathfrak{gl}(n, \mathbb{K})$ *and* $\mathfrak{u}(n)$ *have one-dimensional center:* $\mathfrak{gl}(n, \mathbb{K}) = \mathbb{K} \cdot \mathrm{id} \oplus \mathfrak{sl}(n, \mathbb{K}), \mathfrak{u}(n) = i\mathbb{R} \cdot \mathrm{id} \oplus \mathfrak{su}(n).$ *(As before,* $\mathbb{K}$ *is either* $\mathbb{R}$ *or* $\mathbb{C}$.*)*

Proof. For each of these subalgebras, consider the trace form B_V where V is the defining representation ($\mathbb{K}^n$ for $\mathfrak{gl}(n, \mathbb{K})$, $\mathfrak{sl}(n, \mathbb{K})$, $\mathfrak{so}(n, \mathbb{K})$; $\mathbb{C}^n$ for $\mathfrak{su}(n), \mathfrak{u}(n)$ and $\mathbb{K}^{2n}$ for $\mathfrak{sp}(n, \mathbb{K})$). Then this form is non-degenerate. Indeed, for $\mathfrak{gl}(n)$ it follows because $B(x, y) = \sum x_{ij} y_{ji}$, which is obviously non-degenerate; for $\mathfrak{sl}(n)$ it follows from the result for $\mathfrak{gl}(n)$ and decomposition $\mathfrak{gl}(n) = \mathbb{K} \cdot \mathrm{id} \oplus \mathfrak{sl}(n)$, with the two summands being orthogonal with respect to the form B.

For $\mathfrak{so}(n)$, we have $B(x, y) = \sum x_{ij} y_{ji} = -2 \sum_{i>j} x_{ij} y_{ij}$ so it is again non-degenerate. Similarly, for $\mathfrak{u}(n)$ we have $B(x, y) = -\operatorname{tr} x \bar{y}^t = -\sum x_{ij} \bar{y}_{ij}$; in particular, $B(x, x) = -\sum |x_{ij}|^2$, so this form is negative definite and in particular, non-degenerate. Therefore, its restriction to $\mathfrak{su}(n) \subset \mathfrak{u}(n)$ is also negative definite and thus non-degenerate.

The non-degeneracy of this form for $\mathfrak{sp}(n, \mathbb{K})$ is left as an exercise (Exercise 5.4).

Thus, by Theorem 5.48 we see that each of these Lie algebras is reductive. Since the center of each of them is easy to compute (see Example 4.24), we get the statement of the theorem. $\qquad\square$

5.8. Killing form and Cartan's criterion

In the previous section, we have shown that for any representation V of a Lie algebra $\mathfrak{g}$, the bilinear form $B_V(x, y) = \operatorname{tr}(\rho(x)\rho(y))$ is symmetric and invariant. An important special case is when we take V to be the adjoint representation.

Definition 5.50. The Killing form is the bilinear form on $\mathfrak{g}$ defined by $K(x, y) = \operatorname{tr}(\operatorname{ad} x \operatorname{ad} y).$

The notation $K(x, y)$ can be ambiguous: if we have a subalgebra $\mathfrak{h} \subset \mathfrak{g}$, then $K(x, y), x, y \in \mathfrak{h}$, can mean either trace in $\mathfrak{g}$ or trace in $\mathfrak{h}$. In such cases we will write $K^{\mathfrak{h}}$ for Killing form of $\mathfrak{h}$ and $K^{\mathfrak{g}}$ for the restriction of Killing form of $\mathfrak{g}$ to $\mathfrak{h}$. Note, however, that if I is an ideal in $\mathfrak{g}$, then K^I coincides with the restriction of $K^{\mathfrak{g}}$ to I (see Exercise 5.1).

It follows from Proposition 5.47 that the Killing form is a symmetric invariant form on $\mathfrak{g}$.

Example 5.51. Let $\mathfrak{g} = \mathfrak{sl}(2, \mathbb{C})$. Then in the basis e, h, f, the operators ad e, ad h, ad f are given by

$$\text{ad } e = \begin{pmatrix} 0 & -2 & 0 \\ 0 & 0 & 1 \\ 0 & 0 & 0 \end{pmatrix}, \quad \text{ad } h = \begin{pmatrix} 2 & 0 & 0 \\ 0 & 0 & 0 \\ 0 & 0 & -2 \end{pmatrix}, \quad \text{ad } f = \begin{pmatrix} 0 & 0 & 0 \\ -1 & 0 & 0 \\ 0 & 2 & 0 \end{pmatrix}$$

so an explicit computation shows that the Killing form is given by K(h,h)=8, $K(e,f) = K(f,e) = 4$, and $K(h,e) = K(h,f) = 0$. Thus, $K(x,y) = 4\,\text{tr}(xy)$. This is not surprising: we already know that $\mathfrak{sl}(2, \mathbb{C})$ is simple, and by Exercise 4.5, this implies that the invariant bilinear form, if exists, is unique up to a factor.

The following two theorems show that non-degeneracy of Killing form is closely related to semisimplicity of $\mathfrak{g}$.

Theorem 5.52 (Cartan's criterion of solvability). *Lie algebra $\mathfrak{g}$ is solvable iff $K([\mathfrak{g}, \mathfrak{g}], \mathfrak{g}) = 0$, i.e. $K(x, y) = 0$ for any $x \in [\mathfrak{g}, \mathfrak{g}], y \in \mathfrak{g}$.*

Theorem 5.53 (Cartan's criterion of semisimplicity). *Lie algebra is semisimple iff the Killing form is non-degenerate.*

The proof of these theorems is based on Jordan decomposition, i.e. the decomposition of a linear operator in a sum of a semisimple (which, for operators in finite-dimensional complex vector spaces, is the same as diagonalizable) and nilpotent ones. We state here some results about this decomposition. Their proof, which is pure linear algebra, is given in Section 5.9.

Theorem 5.54. *Let V be a finite-dimensional complex vector space.*

(1) *Any linear operator A can be uniquely written as a sum of commuting semisimple and nilpotent operators:*

$$A = A_s + A_n, \quad A_s A_n = A_n A_s, \quad A_n \text{ nilpotent}, \quad A_s \text{ semisimple} \quad (5.15)$$

(2) *For an operator $A: V \to V$, define ad $A:$ End$(V) \to$ End(V) by ad $A.B = AB - BA$. Then*

$$(\text{ad } A)_s = \text{ad } A_s$$

and ad A_s can be written in the form ad $A_s = P(\text{ad } A)$ for some polynomial $P \in t\mathbb{C}[t]$ (depending on A).

(3) *Define $\overline{A}_s$ to be the operator which has the same eigenspaces as A_s but complex conjugate eigenvalues: if $A_s v = \lambda v$, then $\overline{A}_s v = \overline{\lambda} v$. Then $\operatorname{ad} \overline{A}_s$ can be written in the form $\operatorname{ad} \overline{A}_s = Q(\operatorname{ad} A)$ for some polynomial $Q \in t\mathbb{C}[t]$ (depending on A).*

Using this theorem, we can now give the proof of Cartan's criterion.

Proof of Theorem 5.52. First, note that if $\mathfrak{g}$ is a real Lie algebra, then $\mathfrak{g}$ is solvable iff $\mathfrak{g}_\mathbb{C}$ is solvable (Theorem 5.29), and $K([\mathfrak{g}, \mathfrak{g}], \mathfrak{g}) = 0$ iff $K([\mathfrak{g}_\mathbb{C}, \mathfrak{g}_\mathbb{C}], \mathfrak{g}_\mathbb{C}) = 0$ (obvious). Thus, it suffices to prove the theorem for complex Lie algebras. So from now on we assume that $\mathfrak{g}$ is complex.

Assume that $\mathfrak{g}$ is solvable. Then by Lie's theorem, there is a basis in $\mathfrak{g}$ such that all $\operatorname{ad} x$ are upper-triangular. Then in this basis, operators $\operatorname{ad} y, y \in [\mathfrak{g}, \mathfrak{g}]$ are strictly upper-triangular, so $\operatorname{tr}(\operatorname{ad} x \operatorname{ad} y) = 0$.

To prove the opposite direction, we first prove the following lemma.

Lemma 5.55. *Let V be a complex vector space and $\mathfrak{g} \subset \mathfrak{gl}(V)$ – a Lie subalgebra such that for any $x \in [\mathfrak{g}, \mathfrak{g}], y \in \mathfrak{g}$ we have $\operatorname{tr}(xy) = 0$. Then $\mathfrak{g}$ is solvable.*

Proof. Let $x \in [\mathfrak{g}, \mathfrak{g}]$. By Theorem 5.54, it can be written in the form $x = x_s + x_n$. Consider now $\operatorname{tr}(x\overline{x}_s)$ where $\overline{x}_s$ is as in Theorem 5.54. On one hand, we see that $\operatorname{tr}(x\overline{x}_s) = \sum \lambda_i \overline{\lambda}_i = \sum |\lambda_i|^2$, where λ_i are eigenvalues of x. On the other hand, if $x = \sum [y_i, z_i]$, then

$$\operatorname{tr}(x\overline{x}_s) = \operatorname{tr}\left(\sum [y_i, z_i]\overline{x}_s\right) = \sum \operatorname{tr}(y_i[z_i, \overline{x}_s]) = -\sum \operatorname{tr}(y_i[\overline{x}_s, z_i]).$$

By Theorem 5.54, $[\overline{x}_s, z_i] = \operatorname{ad} \overline{x}_s.z_i = Q(\operatorname{ad} x).z_i \in [\mathfrak{g}, \mathfrak{g}]$. Thus by assumption $\operatorname{tr}(x\overline{x}_s) = 0$. On the other hand, $\operatorname{tr}(x\overline{x}_s) = \sum |\lambda_i|^2$. Therefore, all eigenvalues of x are zero and x is nilpotent. By one of the versions of Engel's theorem (Theorem 5.33), this implies that $[\mathfrak{g}, \mathfrak{g}]$ is nilpotent, so $\mathfrak{g}$ is solvable. This completes the proof of Lemma 5.55. □

Now the proof of Theorem 5.52 easily follows. Indeed, if $K(\mathfrak{g}, [\mathfrak{g}, \mathfrak{g}]) = 0$, then by Lemma 5.55, $\operatorname{ad}(\mathfrak{g}) \subset \mathfrak{gl}(\mathfrak{g})$ is solvable. Thus, both $\mathfrak{z}(\mathfrak{g})$, and $\mathfrak{g}/\mathfrak{z}(\mathfrak{g}) = \operatorname{ad}(\mathfrak{g})$ are solvable. By Theorem 5.29, this implies that $\mathfrak{g}$ is solvable. □

Proof of Theorem 5.53. If K is non-degenerate, then by Theorem 5.48, $\mathfrak{g}$ is reductive. On the other hand, if $x \in \mathfrak{z}(\mathfrak{g})$, then $\operatorname{ad} x = 0$, so $x \in \operatorname{Ker} K$. Thus, $\mathfrak{z}(\mathfrak{g}) = 0$, so $\mathfrak{g}$ is semisimple.

Conversely, assume that $\mathfrak{g}$ is semisimple. Consider $I = \operatorname{Ker} K$; by Lemma 5.45, I is an ideal in $\mathfrak{g}$. Since restriction of K to I coincides with the Killing form of I (Exercise 5.1), the Killing form of I is zero and thus, by previous theorem, I is solvable. But $\mathfrak{g}$ is semisimple, so $I = 0$. Thus, K is non-degenerate.   □

5.9. Jordan decomposition

In this section, we give the proof of the Jordan decomposition for linear operators, which was used in Section 5.8, and several related results.

Throughout this section, V is a finite-dimensional complex vector space.

Definition 5.56. An operator $A\colon V \to V$ is called *nilpotent* if $A^n = 0$ for sufficiently large n.

An operator $A\colon V \to V$ is called *semisimple* if any A-invariant subspace has an A-invariant complement: if $W \subset V$, $AW \subset W$, then there exists $W' \subset V$ such that $V = W \oplus W'$, $AW' \subset W'$.

Lemma 5.57.

(1) *An operator $A\colon V \to V$ is semisimple iff it is diagonalizable.*
(2) *Let $A\colon V \to V$ be semisimple, and $W \subset V$ stable under A: $AW \subset W$. Then restrictions of A to W and to V/W are semisimple operators.*
(3) *Sum of two commuting semisimple operators is semisimple. Sum of two commuting nilpotent operators is nilpotent.*

Proof. If A is semisimple, let v_1 be an eigenvector of A; then $V = \mathbb{C}v_1 \oplus W$ for some A-invariant subspace W. Note let v_2 be an eigenvector for $A|_W$; repeating in this way, we get an eigenbasis for A. Conversely, suppose that A is diagonalizable; then one can write $V = \bigoplus V_{\lambda_i}$, where λ_i are distinct eigenvalues of A and V_{λ_i} is the corresponding eigenspace. Then any A-invariant subspace W also splits into direct sum: $W = \bigoplus (W \cap V_{\lambda_i})$. Indeed, if $p_i \in \mathbb{C}[t]$ is the polynomial such that $p(\lambda_i) = 1, p(\lambda_j) = 0$ for $i \neq j$, then $p_i(A)$ is the projector $V \to V_{\lambda_i}$, and thus any vector $w \in W$ can be written as $w = \sum_i w_i, w_i = p_i(A)w \in W \cap V_{\lambda_i}$. Therefore, $W = \bigoplus W_i, W_i = W \cap V_{\lambda_i}$. Choosing in each V_{λ_i} a subspace $W_i^\perp$ so that $V_{\lambda_i} = W_i \oplus W_i^\perp$, we see that $V = W \oplus W^\perp$, where $W^\perp = \oplus W_i^\perp$.

The same argument also shows that if $W \subset V$ is A-invariant, then $A|_W$ is diagonalizable and therefore semisimple, thus proving the second part.

Finally, the last part is a well-known result of linear algebra.   □

Remark 5.58. Over $\mathbb{R}$, not every semisimple operator is diagonalizable; however, the second and third parts of the theorem remain true over $\mathbb{R}$.

Theorem 5.59. *Any linear operator* $A \colon V \to V$ *can be uniquely written as a sum of commuting semisimple and nilpotent operators:*

$$A = A_s + A_n, \quad A_s A_n = A_n A_s, \quad A_n \text{ nilpotent}, \quad A_s \text{ semisimple} \qquad (5.16)$$

Moreover, A_s, A_n *can be written as polynomials of* A: $A_s = p(A), A_n = A - p(A)$ *for some polynomial* $p \in \mathbb{C}[t]$ *depending on* A.

Decomposition (5.16) is called the Jordan decomposition of A.

Proof. It is well-known from linear algebra that one can decompose V in the direct sum of generalized eigenspaces: $V = \bigoplus V_{(\lambda)}$, where λ runs over the set of distinct eigenvalues of A and $V_{(\lambda)}$ is the generalized eigenspace with eigenvalue λ: restriction of $A - \lambda$ id to $V_{(\lambda)}$ is nilpotent.

Define A_s by $A_s|_{V_{(\lambda)}} = \lambda$ id, and A_n by $A_n = A - A_s$. Then it is immediate from the definition that A_s is semisimple and A_n is nilpotent. It is also easy to see that they commute: in fact, A_s commutes with any operator $V_{(\lambda)} \to V_{(\lambda)}$. This shows existence of Jordan decomposition.

Let us also show that so defined A_s, A_n can be written as polynomials in A. Indeed, let $p \in \mathbb{C}[t]$ be defined by system of congruences

$$p(t) \equiv \lambda_i \quad \mod (t - \lambda_i)^{n_i}$$

where λ_i are distinct eigenvalues of A and $n_i = \dim V_{(\lambda_i)}$. By the Chinese remainder theorem, such a polynomial exists. Since $(A - \lambda_i)^{n_i} = 0$ on $V_{(\lambda_i)}$, we see that $p(A)|_{V_{(\lambda)}} = \lambda = A_s|_{V_{(\lambda)}}$. Thus, $A_s = p(A)$.

Finally, let us prove uniqueness. Let A_s, A_n be as defined above. Assume that $A = A'_s + A'_n$ is another Jordan decomposition. Then $A_s + A_n = A'_s + A'_n$. Since A'_s, A'_n commute with each other, they commute with A; since $A_s = p(A)$, we see that A_s, A_n commute with A'_s, A'_n. Consider now the operator $A_s - A'_s = A'_n - A_n$. On one hand, it is semisimple as a difference of two commuting semisimple operators. On the other hand, it is nilpotent as a difference of two commuting nilpotent operators (see Lemma 5.57). Thus, all its eigenvalues are zero; since it is semisimple, it shows that it is a zero operator, so $A_s = A'_s, A_n = A'_n$. $\square$

The proof also shows that it is possible to choose a basis in V such that in this basis, A_s is diagonal and A_n is strictly upper-triangular, and that if 0 is an eigenvalue of A, then $p(0) = 0$.

Theorem 5.60. *Let* A *be an operator* $V \to V$. *Define* $\operatorname{ad} A \colon \operatorname{End}(V) \to \operatorname{End}(V)$ *by* $\operatorname{ad} A.B = AB - BA$. *Then* $(\operatorname{ad} A)_s = \operatorname{ad} A_s$, *and* $\operatorname{ad} A_s$ *can be written in the form* $\operatorname{ad} A_s = P(\operatorname{ad} A)$ *for some polynomial* $P \in \mathbb{C}[t]$ *such that* $P(0) = 0$.

Proof. Let $A = A_s + A_n$ be the Jordan decomposition for A. Then $\operatorname{ad} A = \operatorname{ad} A_s + \operatorname{ad} A_n$, and it is immediate to check that $\operatorname{ad} A_s$, $\operatorname{ad} A_n$ commute.

Choose a basis in V such that in this basis, A_s is diagonal, A_n is strictly upper-triangular. Then it also gives a basis of matrix units E_{ij} in $\operatorname{End}(V)$. In this basis, the action of $\operatorname{ad} A_s$ is diagonal: $\operatorname{ad} A_s.E_{ij} = (\lambda_i - \lambda_j)E_{ij}$, as is easily verified by a direct computation. Using this basis, it is also easy to check that $\operatorname{ad} A_n$ is nilpotent (see Exercise 5.7). Thus, $\operatorname{ad} A = \operatorname{ad} A_s + \operatorname{ad} A_n$ is the Jordan decomposition for $\operatorname{ad} A$, so $(\operatorname{ad} A)_s = \operatorname{ad} A_s$.

By Theorem 5.59 applied to operator $\operatorname{ad} A$, we see that $(\operatorname{ad} A)_s$ can be written in the form $(\operatorname{ad} A)_s = P(\operatorname{ad} A)$ for some polynomial $P \in \mathbb{C}[t]$; moreover, since 0 is an eigenvalue of $\operatorname{ad} A$ (e.g., $\operatorname{ad} A.A = 0$), we see that $P(0) = 0$. $\qquad\square$

Theorem 5.61. *Let A be an operator $V \to V$. Define $\overline{A}_s$ to be the operator which has the same eigenspaces as A_s but complex conjugate eigenvalues: if $A_s v = \lambda v$, then $\overline{A}_s v = \bar{\lambda} v$. Then $\operatorname{ad} \overline{A}_s$ can be written in the form $\operatorname{ad} \overline{A}_s = Q(\operatorname{ad} A)$ for some polynomial $Q \in t\mathbb{C}[t]$ (depending on A).*

Proof. Let $\{v_i\}$ be a basis of eigenvectors for A_s: $A_s v_i = \lambda_i v_i$ so that $\overline{A}_s v_i = \bar{\lambda}_i v_i$. Let E_{ij} be the corresponding basis in $\operatorname{End}(V)$; then, as discussed in the proof of Theorem 5.60, in this basis $\operatorname{ad} A_s$ is given by $\operatorname{ad} A_s.E_{ij} = (\lambda_i - \lambda_j)E_{ij}$, and $\operatorname{ad} \overline{A}_s.E_{ij} = (\bar{\lambda}_i - \bar{\lambda}_j)E_{ij}$.

Choose a polynomial $f \in \mathbb{C}[t]$ such that $f(\lambda_i - \lambda_j) = \bar{\lambda}_i - \bar{\lambda}_j$ (in particular, $f(0) = 0$); such a polynomial exists by interpolation theorem. Then $\operatorname{ad} \overline{A}_s = f(\operatorname{ad} A_s) = f(P(\operatorname{ad} A))$ where P is as in Theorem 5.60. $\qquad\square$

5.10. Exercises

5.1.

(1) Let V be a representation of $\mathfrak{g}$ and $W \subset V$ be a subrepresentation. Then $B_V = B_W + B_{V/W}$, where B_V is defined by (5.14).

(2) Let $I \subset \mathfrak{g}$ be an ideal. Then the restriction of the Killing form of $\mathfrak{g}$ to I coincides with the Killing form of I.

5.2. Show that for $\mathfrak{g} = \mathfrak{sl}(n, \mathbb{C})$, the Killing form is given by $K(x, y) = 2n \operatorname{tr}(xy)$.

5.3. Let $\mathfrak{g} \subset \mathfrak{gl}(n, \mathbb{C})$ be the subspace consisting of block-triangular matrices:

$$\mathfrak{g} = \left\{ \begin{pmatrix} A & B \\ 0 & D \end{pmatrix} \right\}$$

where A is a $k \times k$ matrix, B is a $k \times (n-k)$ matrix, and D is a $(n-k) \times (n-k)$ matrix.

(1) Show that $\mathfrak{g}$ is a Lie subalgebra (this is a special case of so-called *parabolic subalgebras*).

(2) Show that radical of $\mathfrak{g}$ consists of matrices of the form $\begin{pmatrix} \lambda \cdot I & B \\ 0 & \mu \cdot I \end{pmatrix}$, and describe $\mathfrak{g}/\operatorname{rad}(\mathfrak{g})$.

5.4. Show that the bilinear form $\operatorname{tr}(xy)$ on $\mathfrak{sp}(n, \mathbb{K})$ is non-degenerate.

5.5. Let $\mathfrak{g}$ be a real Lie algebra with a positive definite Killing form. Show that then $\mathfrak{g} = 0$. [Hint: $\mathfrak{g} \subset \mathfrak{so}(\mathfrak{g})$.]

5.6. Let $\mathfrak{g}$ be a simple Lie algebra.

(1) Show that the invariant bilinear form is unique up to a factor. [Hint: use Exercise 4.5.]

(2) Show that $\mathfrak{g} \simeq \mathfrak{g}^*$ as representations of $\mathfrak{g}$.

5.7. Let V be a finite-dimensional complex vector space and let $A \colon V \to V$ be a strictly upper-triangular operator. Let $F^k \subset \operatorname{End}(V)$, $-n \leq k \leq n$ be the subspace spanned by matrix units E_{ij} with $i - j \leq k$. Show that then $\operatorname{ad} A.F^k \subset F^{k-1}$ and thus, $\operatorname{ad} A \colon \operatorname{End}(V) \to \operatorname{End}(V)$ is nilpotent.

6

Complex semisimple Lie algebras

In this chapter, we begin the study of semisimple Lie algebras and their representations. This is one of the highest achievements of the theory of Lie algebras, which has numerous applications (for example, to physics), not to mention that it is also one of the most beautiful areas of mathematics.

Throughout this chapter, $\mathfrak{g}$ is a finite-dimensional semisimple Lie algebra (see Definition 5.35); unless specified otherwise, $\mathfrak{g}$ is complex.

6.1. Properties of semisimple Lie algebras

Cartan's criterion of semimplicity, proved in Section 5.8, is not very convenient for practical computations. However, it is extremely useful for theoretical considerations.

Proposition 6.1. *Let $\mathfrak{g}$ be a real Lie algebra and $\mathfrak{g}_\mathbb{C}$ – its complexification (see Definition 3.49). Then $\mathfrak{g}$ is semisimple iff $\mathfrak{g}_\mathbb{C}$ is semisimple.*

Proof. Immediately follows from Cartan's criterion of semisimplicity. □

Remark 6.2. This theorem fails if we replace the word "semisimple" by "simple": there exist simple real Lie algebras $\mathfrak{g}$ such that $\mathfrak{g}_\mathbb{C}$ is a direct sum of two simple algebras.

Theorem 6.3. *Let $\mathfrak{g}$ be a semisimple Lie algebra, and $I \subset \mathfrak{g}$ – an ideal. Then there is an ideal I' such that $\mathfrak{g} = I \oplus I'$.*

Proof. Let $I^\perp$ be the orthogonal complement with respect to the Killing form. By Lemma 5.45, $I^\perp$ is an ideal. Consider the intersection $I \cap I^\perp$. It is an ideal in $\mathfrak{g}$ with zero Killing form (by Exercise 5.1). Thus, by Cartan criterion, it is solvable. By definition of a semisimple Lie algebra, this means that $I \cap I^\perp = 0$, so $\mathfrak{g} = I \oplus I^\perp$. □

Corollary 6.4. *A Lie algebra is semisimple iff it is a direct sum of simple Lie algebras.*

Proof. Any simple Lie algebra is semisimple by Lemma 5.37, and it is immediate from Cartan criterion that the direct sum of semisimple Lie algebras is semisimple. This proves one direction.

The opposite direction – that each semisimple algebra is a direct sum of simple ones – easily follows by induction from the previous theorem. $\square$

Corollary 6.5. *If $\mathfrak{g}$ is a semisimple Lie algebra, then $[\mathfrak{g}, \mathfrak{g}] = \mathfrak{g}$.*

Indeed, for a simple Lie algebra it is clear because $[\mathfrak{g}, \mathfrak{g}]$ is an ideal in $\mathfrak{g}$ which can not be zero (otherwise, $\mathfrak{g}$ would be abelian).

Proposition 6.6. *Let $\mathfrak{g} = \mathfrak{g}_1 \oplus \cdots \oplus \mathfrak{g}_k$ be a semisimple Lie algebra, with $\mathfrak{g}_i$ being simple. Then any ideal I in $\mathfrak{g}$ is of the form $I = \bigoplus_{i \in J} \mathfrak{g}_i$ for some subset $J \subset \{1, \ldots, k\}$.*

Note that it is not an "up to isomorphism" statement: I is not just isomorphic to sum of some of $\mathfrak{g}_i$ but actually equal to such a sum as a subspace in $\mathfrak{g}$.

Proof. The proof goes by induction in k. Let $\pi_k \colon \mathfrak{g} \to \mathfrak{g}_k$ be the projection. Consider $\pi_k(I) \subset \mathfrak{g}_k$. Since $\mathfrak{g}_k$ is simple, either $\pi_k(I) = 0$, in which case $I \subset \mathfrak{g}_1 \oplus \cdots \oplus \mathfrak{g}_{k-1}$ and we can use induction assumption, or $\pi_k(I) = \mathfrak{g}_k$. Then $[\mathfrak{g}_k, I] = [\mathfrak{g}_k, \pi_k(I)] = \mathfrak{g}_k$. Since I is an ideal, $I \supset \mathfrak{g}_k$, so $I = I' \oplus \mathfrak{g}_k$ for some subspace $I' \subset \mathfrak{g}_1 \oplus \cdots \oplus \mathfrak{g}_{k-1}$. It is immediate that then I' is an ideal in $\mathfrak{g}_1 \oplus \cdots \oplus \mathfrak{g}_{k-1}$ and the result again follows from the induction assumption. $\square$

Corollary 6.7. *Any ideal in a semisimple Lie algebra is semisimple. Also, any quotient of a semisimple Lie algebra is semisimple.*

Finally, recall that we have denoted by Der $\mathfrak{g}$ the Lie algebra of all derivations of $\mathfrak{g}$ (see (3.14)) and by Aut $\mathfrak{g}$ the group of all automorphisms of $\mathfrak{g}$ (see Example 3.33).

Proposition 6.8. *If $\mathfrak{g}$ is a semisimple Lie algebra, and G–a connected Lie group with Lie algebra $\mathfrak{g}$, then Der $\mathfrak{g} = \mathfrak{g}$, and Aut $\mathfrak{g}/ \operatorname{Ad} G$ is discrete, where $\operatorname{Ad} G = G/Z(G)$ is the adjoint group associated with G (see (3.15)).*

Proof. Recall that for any $x \in \mathfrak{g}$, $\operatorname{ad} x \colon \mathfrak{g} \to \mathfrak{g}$ is a derivation. This gives a natural morphism of Lie algebras $\mathfrak{g} \to$ Der $\mathfrak{g}$. Since the center $\mathfrak{z}(\mathfrak{g}) = 0$, this morphism is injective, so $\mathfrak{g}$ is a subalgebra in Der $\mathfrak{g}$.

The definition of the derivation immediately shows that for any derivation δ and $x \in \mathfrak{g}$, we have $\operatorname{ad}(\delta(x)) = [\delta, \operatorname{ad} x]$ as operators $\mathfrak{g} \to \mathfrak{g}$. Thus, $\mathfrak{g} \subset$ Der $\mathfrak{g}$ is an ideal.

Let us now extend the Killing form of $\mathfrak{g}$ to Der $\mathfrak{g}$ by letting $K(\delta_1, \delta_2) = \mathrm{tr}_{\mathfrak{g}}(\delta_1\delta_2)$ and consider the orthogonal complement $I = \mathfrak{g}^{\perp} \subset$ Der $\mathfrak{g}$. Since K is invariant, I is an ideal; since restriction of K to $\mathfrak{g}$ is non-degenerate, $I \cap \mathfrak{g} = 0$. Thus, Der $\mathfrak{g} = \mathfrak{g} \oplus I$; since both $\mathfrak{g}, I$ are ideals, we have $[I, \mathfrak{g}] = 0$, which implies that for every $\delta \in I, x \in \mathfrak{g}$, we have $\mathrm{ad}(\delta(x)) = [\delta, \mathrm{ad}\, x] = 0$, so $\delta(x) = 0$. Thus, $I = 0$.

Since Aut $\mathfrak{g}$ is a Lie group with Lie algebra Der $\mathfrak{g}$ (see Example 3.33), the second statement of the theorem immediately follows from the first one. $\square$

6.2. Relation with compact groups

In Section 5.8, we have shown that the Killing form of $\mathfrak{g}$ is non-degenerate if and only if $\mathfrak{g}$ is semisimple. However, in the case of *real* $\mathfrak{g}$, one might also ask whether the Killing form is positive definite, negative definite, or neither. More generally, the same question can be asked about the trace form in any representation: $B_V(x, y) = \mathrm{tr}_V(xy)$.

It turns out that the answer to this question is closely related to the question of compactness of the corresponding Lie group.

Example 6.9. Let $\mathfrak{g} = \mathfrak{u}(n)$ be the Lie algebra of the unitary group, i.e. the Lie algebra of skew-Hermitian matrices. Then the form $(x, y) = \mathrm{tr}(xy)$ is negative definite.

Indeed, $\mathrm{tr}(xy) = -\mathrm{tr}(x\bar{y}^t)$, and $\mathrm{tr}(x^2) = -\mathrm{tr}(x\bar{x}^t) = -\sum |x_{ij}|^2 \leq 0$, with equality only for $x = 0$.

Theorem 6.10. *Let G be a compact real Lie group. Then $\mathfrak{g} = \mathrm{Lie}(G)$ is reductive, and the Killing form of $\mathfrak{g}$ is negative semidefinite, with $\mathrm{Ker}\, K = \mathfrak{z}(\mathfrak{g})$ (the center of $\mathfrak{g}$); the Killing form of the semisimple part $\mathfrak{g}/\mathfrak{z}(\mathfrak{g})$ is negative definite.*

Conversely, let $\mathfrak{g}$ be a semisimple real Lie algebra with a negative definite Killing form. Then $\mathfrak{g}$ is a Lie algebra of a compact real Lie group.

Proof. If G is compact, then by Theorem 4.40, every complex representation $\rho\colon G \to \mathrm{GL}(V)$ of G is unitary, so $\rho(G) \subset U(V), \rho(\mathfrak{g}) \subset \mathfrak{u}(V)$ (where $U(V)$ is the group of unitary operators $V \to V$). By Example 6.9, this implies that the trace form $B_V(x, y)$ is negative semidefinite, with $\mathrm{Ker}\, B_V = \mathrm{Ker}\, \rho$.

Applying this to the complexified adjoint representation $V = \mathfrak{g}_{\mathbb{C}}$, we see that the Killing form is negative semidefinite, with $\mathrm{Ker}\, K = \mathfrak{z}(\mathfrak{g})$.

Conversely, assume that $\mathfrak{g}$ is a real Lie algebra with negative definite Killing form K. Let G be a connected Lie group with Lie algebra $\mathfrak{g}$. Then $B(x, y) = -K(x, y)$ is positive definite and Ad G invariant. This shows that $\mathrm{Ad}(G) \subset \mathrm{SO}(\mathfrak{g})$ (the orthogonal group). Since $\mathrm{Ad}(G)$ is the connected component of unity of the group Aut $\mathfrak{g}$ (see Proposition 6.8), and Aut $\mathfrak{g} \subset \mathrm{GL}(\mathfrak{g})$ is a closed Lie subgroup (see Example 3.33), $\mathrm{Ad}(G)$ is a closed Lie subgroup in the compact group $\mathrm{SO}(\mathfrak{g})$. Thus, $\mathrm{Ad}(G)$ is a compact Lie group. Since $\mathrm{Ad}(G) = G/Z(G)$, we have $\mathrm{Lie}(\mathrm{Ad}(G)) = \mathfrak{g}/\mathfrak{z}(\mathfrak{g}) = \mathfrak{g}$, which proves the theorem. $\qquad\square$

Remark 6.11. In fact, one can prove a stronger result: if $\mathfrak{g}$ is a real Lie algebra with negative definite Killing form, then any connected Lie group with Lie algebra $\mathfrak{g}$ is compact. In particular, the simply-connected Lie group with Lie algebra $\mathfrak{g}$ is compact.

One might also ask for which real Lie algebras the Killing form is *positive definite*. Unfortunately, it turns out that there are not many such algebras.

Lemma 6.12. *If $\mathfrak{g}$ is a real Lie algebra with a positive definite Killing form, then $\mathfrak{g} = 0$.*

The proof of this lemma is given as an exercise (Exercise 5.5).

Finally, let us discuss the relation between complex semisimple Lie algebras and compact groups. It turns out that the only compact complex Lie groups are tori (see Exercise 3.19). Instead, we could take real compact Lie groups and corresponding Lie algebras, then consider their complexifications. By Theorem 6.10, such complex Lie algebras will be reductive. A natural question is whether any reductive complex Lie algebra can be obtained in this way. The following theorem (which for simplicity is stated only for semisimple Lie algebras) provides the answer.

Theorem 6.13. *Let $\mathfrak{g}$ be a complex semisimple Lie algebra. Then there exists a real subalgebra $\mathfrak{k}$ such that $\mathfrak{g} = \mathfrak{k} \otimes \mathbb{C}$ and $\mathfrak{k}$ is a Lie algebra of a compact Lie group K. The Lie algebra $\mathfrak{k}$ is called the* compact real form *of $\mathfrak{g}$; it is unique up to conjugation.*

If G is a connected complex Lie group with Lie algebra $\mathfrak{g}$, then the compact group K can be chosen so that $K \subset G$. In this case, K is called the compact real form of the Lie group G.

The proof of this theorem will not be given here. Interested readers can find a discussion of this theorem in [15, Section I.7] or in [32].

Example 6.14. For $\mathfrak{g} = \mathfrak{sl}(n, \mathbb{C}), G = \mathrm{SL}(n, \mathbb{C})$, the compact form is $\mathfrak{k} = \mathfrak{su}(n), K = \mathrm{SU}(n)$.

6.3. Complete reducibility of representations

In this section, we will show one of fundamental results of the theory of semisimple Lie algebras: every representation of a semisimple Lie algebra is completely reducible. Throughout this section, $\mathfrak{g}$ is a semisimple complex Lie algebra and V – a finite-dimensional complex representation of $\mathfrak{g}$.

This result can be proved in several ways. Historically, the first proof of this result was given by H. Weyl using the theory of compact groups. Namely, if $\mathfrak{g}$ is a semisimple complex Lie algebra, then by Theorem 6.13 $\mathfrak{g}$ can be written as a complexification of a real Lie algebra $\mathfrak{k} = \mathrm{Lie}(K)$ for some compact, connected simply-connected group K. Then complex representations of $\mathfrak{g}$, $\mathfrak{k}$ and K are the same, and by Theorem 4.40, every representation of K is completely reducible. This argument is commonly called "Weyl's unitary trick".

However, there is a completely algebraic proof of complete reducibility. It uses some basic homological algebra: obstruction to complete reducibility is described by a certain type of cohomology, and we will show that this cohomology vanishes. To do so, we will use a special central element in the universal enveloping algebra, called the *Casimir element*.

Proposition 6.15. *Let $\mathfrak{g}$ be a Lie algebra, and B – a non-degenerate invariant symmetric bilinear form on $\mathfrak{g}$. Let x_i be a basis of $\mathfrak{g}$, and x^i – the dual basis with respect to B. Then the element*

$$C_B = \sum x_i x^i \in U\mathfrak{g}$$

does not depend on the choice of the basis x_i and is central. It is called the Casimir element determined by form B.

In particular, if $\mathfrak{g}$ is semisimple and K is the Killing form, then the element $C_K \in U\mathfrak{g}$ is called simply the Casimir element.

Proof. Independence of choice of basis follows from the fact that the element $I = \sum x_i \otimes x^i \in \mathfrak{g} \otimes \mathfrak{g}$ is independent of the choice of basis: under the identification $\mathfrak{g} \otimes \mathfrak{g} \simeq \mathfrak{g} \otimes \mathfrak{g}^* = \mathrm{End}(\mathfrak{g})$ given by the form B, this element becomes the identity operator in $\mathrm{End}(\mathfrak{g})$.

This also shows that $I = \sum x_i \otimes x^i$ is ad $\mathfrak{g}$–invariant: indeed, identification $\mathfrak{g} \otimes \mathfrak{g} \simeq \mathfrak{g} \otimes \mathfrak{g}^* = \mathrm{End}(\mathfrak{g})$ is a morphism of representations, and the identity operator commutes with the action of $\mathfrak{g}$ and thus is ad $\mathfrak{g}$ invariant.

Since the multiplication map $\mathfrak{g} \otimes \mathfrak{g} \to U\mathfrak{g}$ is a morphism of representations, we see that $C_B = \sum x_i x^i$ is ad $\mathfrak{g}$–invariant and thus central (see Proposition 5.7). $\qquad\square$

Example 6.16. Let $\mathfrak{g} = \mathfrak{sl}(2, \mathbb{C})$ with the bilinear form defined by $(x, y) = \text{tr}(xy)$. Then the Casimir operator is given by $C = \frac{1}{2}h^2 + fe + ef$ (compare with Example 5.6, where centrality of C was shown by a direct computation).

Remark 6.17. Note that if $\mathfrak{g}$ is simple, then by Exercise 4.5, the invariant bilinear form is unique up to a constant: any such form is a multiple of the Killing form. Thus, in this case the Casimir element is also unique up to a constant.

Proposition 6.18. *Let* V *be a non-trivial irreducible representation of a semisimple Lie algebra* $\mathfrak{g}$. *Then there exists a central element* $C_V \in Z(U\mathfrak{g})$ *which acts by a non-zero constant in* V *and which acts by zero in the trivial representation.*

Proof. Let $B_V(x, y) = \text{tr}_V(\rho(x)\rho(y))$; by Proposition 5.47, this form is an invariant bilinear form. If B_V is non-degenerate, then let $C_V = C_{B_V}$ be the Casimir element of $\mathfrak{g}$ defined by form B_V. Obviously, C_V acts by zero in $\mathbb{C}$. Since V is irreducible, by Schur lemma C_V acts in V by a constant: $C_V = \lambda \, \text{id}_V$. On the other hand, $\text{tr}(C_V) = \sum \text{tr}(x_i x^i) = \dim \mathfrak{g}$, because by definition of $B, \text{tr}(x_i x^i) = B(x_i, x^i) = 1$. Thus, $\lambda = \frac{\dim \mathfrak{g}}{\dim V} \neq 0$, which proves the proposition in this special case.

In general, let $I = \text{Ker } B_V \subset \mathfrak{g}$. Then it is an ideal in $\mathfrak{g}$, and $I \neq \mathfrak{g}$ (otherwise, by Lemma 5.55, $\rho(\mathfrak{g}) \subset \mathfrak{gl}(V)$ is solvable, which is impossible as it is a quotient of a semisimple Lie algebra and thus itself semisimple). By results of Theorem 6.3, $\mathfrak{g} = I \oplus \mathfrak{g}'$ for some non-zero ideal $\mathfrak{g}' \subset \mathfrak{g}$. By Corollary 6.7, $\mathfrak{g}'$ is semisimple, and restriction of B_V to $\mathfrak{g}'$ is non-degenerate. Let C_V be the Casimir element of $\mathfrak{g}'$ corresponding to the form B_V. Since $I, \mathfrak{g}'$ commute, C_V will be central in $U\mathfrak{g}$, and the same argument as before shows that it acts in V by $\frac{\dim \mathfrak{g}'}{\dim V} \neq 0$, which completes the proof. $\qquad\square$

Remark 6.19. In fact, a stronger result is known: if we let C be the Casimir element defined by the Killing form, then C acts by a non-zero constant in any nontrivial irreducible representation. However, this is slightly more difficult to prove.

Now we are ready to prove the main result of this section.

Theorem 6.20. *Any complex finite-dimensional representation of a semisimple Lie algebra* $\mathfrak{g}$ *is completely reducible.*

Proof. The proof assumes some familiarity with basic homological algebra, such as the notion of functors $\text{Ext}^i(V, W)$. We refer the reader to [21] for an overview. They are defined for representations of Lie algebras in the same way as for modules over an associative algebra. In fact, the same definition works for any abelian category, i.e. a category where morphisms form abelian groups and where we have the notion of image and kernel of a morphism satisfying the usual properties.

In particular, the standard argument of homological agebra shows that for fixed V_1, V_2 equivalence classes of extensions $0 \to V_1 \to W \to V_2 \to 0$ are in bijection with $\mathrm{Ext}^1(V_2, V_1)$. Thus, our goal is to show that $\mathrm{Ext}^1(V_2, V_1) = 0$ for any two representations V_1, V_2. This will be done in several steps. For convenience, we introduce the notation $H^1(\mathfrak{g}, V) = \mathrm{Ext}^1(\mathbb{C}, V)$.

Lemma 6.21. *For any irreducible representation V, one has $H^1(\mathfrak{g}, V) = 0$.*

Proof. To prove that $\mathrm{Ext}^1(\mathbb{C}, V) = 0$ it suffices to show that every short exact sequence of the form $0 \to V \to W \to \mathbb{C} \to 0$ splits. So let us assume that we have such an exact sequence.

Let us consider separately two cases: V is a non-trivial irreducible representation and $V = \mathbb{C}$.

If V is a non-trivial irreducible representation, consider the Casimir element C_V as defined in Proposition 6.18. Since it acts in $\mathbb{C}$ by zero and in V by a non-zero constant λ, its eigenvalues in W are 0 with multiplicity 1 and λ with multiplicity $\dim V$. Thus, $W = V \oplus W^0$, where W^0 is the eigenspace for C_V with eigenvalue 0 (which must be one-dimensional). Since C_V is central, W^0 is a subrepresentation; since the kernel of the projection $W \to \mathbb{C}$ is V, it gives an isomorphism $W^0 \simeq \mathbb{C}$. Thus, $W \simeq V \oplus \mathbb{C}$.

If $V = \mathbb{C}$ is a trivial representation, so we have an exact sequence $0 \to \mathbb{C} \to W \to \mathbb{C} \to 0$, then W is a two-dimensional representation such that the action of $\rho(x)$ is strictly upper triangular for all $x \in \mathfrak{g}$. Thus, $\rho(\mathfrak{g})$ is nilpotent, so by Corollary 6.7, $\rho(\mathfrak{g}) = 0$. Thus, $W \simeq \mathbb{C} \oplus \mathbb{C}$ as a representation. $\qquad\square$

This lemma provides the crucial step; the rest is simple homological algebra.

Lemma 6.22. $H^1(\mathfrak{g}, V) = 0$ *for any representation V.*

Proof. If we have a short exact sequence of representations $0 \to V_1 \to V \to V_2 \to 0$, then we have a long exact sequence of Ext groups; in particular,

$$\cdots \to H^1(\mathfrak{g}, V_1) \to H^1(\mathfrak{g}, V) \to H^1(\mathfrak{g}, V_2) \to \cdots$$

Thus, if $H^1(\mathfrak{g}, V_1) = H^1(\mathfrak{g}, V_2) = 0$, then $H^1(\mathfrak{g}, V) = 0$. Since for irreducible representations we have proved that $H^1(\mathfrak{g}, V) = 0$, it is easy to show by induction in dimension that for any representation, $H^1(\mathfrak{g}, V) = 0$. $\qquad\square$

We are now ready to prove Theorem 6.20. Let $0 \to V_1 \to W \to V_2 \to 0$ be a short exact sequence of $\mathfrak{g}$-modules. We need to show that it splits.

Let us apply to this sequence the functor $X \mapsto \mathrm{Hom}_{\mathbb{C}}(V_2, X) = V_2^* \otimes X$ (considered as a $\mathfrak{g}$-module, see Example 4.12). Obviously, this gives a short exact sequence of $\mathfrak{g}$-modules

$$0 \to \mathrm{Hom}_{\mathbb{C}}(V_2, V_1) \to \mathrm{Hom}_{\mathbb{C}}(V_2, W) \to \mathrm{Hom}_{\mathbb{C}}(V_2, V_2) \to 0$$

Now, let us apply to this sequence the functor of $\mathfrak{g}$-invariants: $X \mapsto X^{\mathfrak{g}} = \mathrm{Hom}_{\mathfrak{g}}(\mathbb{C}, X)$. Applying this functor to $\mathrm{Hom}_{\mathbb{C}}(A, B)$ gives $(\mathrm{Hom}_{\mathbb{C}}(A, B))^{\mathfrak{g}} = \mathrm{Hom}_{\mathfrak{g}}(A, B)$ (see Example 4.14).

This functor is left exact but in general not exact, so we get a long exact sequence

$$0 \to \mathrm{Hom}_{\mathfrak{g}}(V_2, V_1) \to \mathrm{Hom}_{\mathfrak{g}}(V_2, W) \to \mathrm{Hom}_{\mathfrak{g}}(V_2, V_2)$$
$$\to \mathrm{Ext}^1(\mathbb{C}, V_2^* \otimes V_1) = H^1(\mathfrak{g}, V_2^* \otimes V_1) \to \dots$$

But since we have already proved that $H^1(\mathfrak{g}, V) = 0$ for any module V, we see that in fact we do have a short exact sequence

$$0 \to \mathrm{Hom}_{\mathfrak{g}}(V_2, V_1) \to \mathrm{Hom}_{\mathfrak{g}}(V_2, W) \to \mathrm{Hom}_{\mathfrak{g}}(V_2, V_2) \to 0$$

In particular, this shows that there exists a morphism $f : V_2 \to W$ which, when composed with projection $W \to V_2$, gives identity morphism $V_2 \to V_2$. This gives a splitting of exact sequence $0 \to V_1 \to W \to V_2 \to 0$. This completes the proof of Theorem 6.20. □

Remark 6.23. The same proof can be rewritten without using the language of Ext groups; see, for example, [46]. This would make it formally accessible to readers with no knowledge of homological algebra. However, this does not change the fact that all arguments are essentially homological in nature; in fact, such a rewriting would obscure the ideas of the proof rather than make them clearer.

The groups $\mathrm{Ext}^1(V, W)$ and in particular, $H^1(\mathfrak{g}, V) = \mathrm{Ext}^1(\mathbb{C}, V)$ used in this proof are just the beginning of a well-developed cohomology theory of Lie algebras. In particular, one can define higher cohomology groups $H^i(\mathfrak{g}, V)$ in a very similar way. The same argument with the Casimir element can be used to show that for a non-trivial irreducible representation V, one has $H^i(\mathfrak{g}, V) = 0$ for $i > 0$. However, it is not true for the trivial representation: $H^i(\mathfrak{g}, \mathbb{C})$ can be non-zero. For example, it can be shown that if G is a connected, simply connected compact real Lie group, and $\mathfrak{g} = \mathrm{Lie}(G)$, then $H^i(\mathfrak{g}, \mathbb{R}) = H^i(G, \mathbb{R})$, where $H^i(G, \mathbb{R})$ are the usual topological cohomology (which can be defined, for example, as De Rham cohomology). We refer the reader to [12] for an introduction to this theory.

Complete reducibility has a number of useful corollaries. One of them is the following result, announced in Section 5.6.

Theorem 6.24. *Any reductive Lie algebra can be written as a direct sum (as a Lie algebra) of semisimple and commutative ideals:*

$$\mathfrak{g} = \mathfrak{z} \oplus \mathfrak{g}_{ss}, \qquad \mathfrak{z} \text{ commutative}, \quad \mathfrak{g}_{ss} \text{ semisimple.}$$

Proof. Consider the adjoint representation of $\mathfrak{g}$. Since the center $\mathfrak{z}(\mathfrak{g})$ acts by zero in an adjoint representation, the adjoint action descends to an action of $\mathfrak{g}' = \mathfrak{g}/\mathfrak{z}(\mathfrak{g})$. By definition of a reductive algebra, $\mathfrak{g}'$ is semisimple. Thus, $\mathfrak{g}$ considered as a representation of $\mathfrak{g}'$ is completely reducible. Since $\mathfrak{z} \subset \mathfrak{g}$ is stable under adjoint action, it is a subrepresentation. By complete reducibility, we can write $\mathfrak{g} = \mathfrak{z} \oplus I$ for some $I \subset \mathfrak{g}$ such that $\mathrm{ad}\, x.I \subset I$ for any $x \in \mathfrak{g}$. Thus, I is an ideal in $\mathfrak{g}$, so $\mathfrak{g} = \mathfrak{z} \oplus I$ as Lie algebras. Obviously, $I \simeq \mathfrak{g}/\mathfrak{z} = \mathfrak{g}'$ is semisimple. $\square$

In a similar way one can prove Levi theorem (Theorem 5.41). We do not give this proof here, referring the reader to [24, 41, 46]. Instead, we just mention that in the language of homological algebra, the Levi theorem follows from the vanishing of cohomology $H^2(\mathfrak{g}, \mathbb{C})$.

6.4. Semisimple elements and toral subalgebras

Recall that the main tool used in the study of representations of $\mathfrak{sl}(2, \mathbb{C})$ in Section 4.8 was the weight decomposition, i.e. decomposing a representation of $\mathfrak{sl}(2, \mathbb{C})$ into direct sum of eigenspaces for h. In order to generalize this idea to other Lie algebras, we need to find a proper analog of h.

Looking closely at the proofs of Section 4.8, we see that the key property of h was the commutation relations $[h, e] = 2e, [h, f] = -2f$ which were used to to show that e, f shift the weight. In other words, $\mathrm{ad}\, h$ is diagonal in the basis e, f, h. This justifies the following definition.

Definition 6.25. An element $x \in \mathfrak{g}$ is called *semisimple* if $\mathrm{ad}\, x$ is a semisimple operator $\mathfrak{g} \to \mathfrak{g}$ (see Definition 5.56).

An element $x \in \mathfrak{g}$ is called *nilpotent* if $\mathrm{ad}\, x$ is a nilpotent operator $\mathfrak{g} \to \mathfrak{g}$.

It is easy to show (see Exercise 6.2) that for $\mathfrak{g} = \mathfrak{gl}(n, \mathbb{C})$ this definition coincides with the usual definition of a semisimple (diagonalizable) operator.

Of course, we do not yet know if such elements exist for any $\mathfrak{g}$. The following theorem, which generalizes Jordan decomposition theorem (Theorem 5.59), answers this question.

Theorem 6.26. *If $\mathfrak{g}$ is a semisimple complex Lie algebra, then any $x \in \mathfrak{g}$ can be uniquely written in the form*

$$x = x_s + x_n,$$

where x_s is semisimple, x_n is nilpotent, and $[x_s, x_n] = 0$. Moreover, if for some $y \in \mathfrak{g}$ we have $[x, y] = 0$, then $[x_s, y] = 0$.

Proof. Uniqueness immediately follows from the uniqueness of the Jordan decomposition for $\operatorname{ad} x$ (Theorem 5.59): if $x = x_s + x_n = x'_s + x'_n$, then $(\operatorname{ad} x)_s = \operatorname{ad} x_s = \operatorname{ad} x'_s$, so $\operatorname{ad}(x_s - x'_s) = 0$. But by definition, a semisimple Lie algebra has zero center, so this implies $x_s - x'_s = 0$.

To prove existence, let us write $\mathfrak{g}$ as direct sum of generalized eigenspaces for $\operatorname{ad} x$: $\mathfrak{g} = \bigoplus \mathfrak{g}_\lambda$, $(\operatorname{ad} x - \lambda \operatorname{id})^n|_{\mathfrak{g}_\lambda} = 0$ for $n \gg 0$.

Lemma 6.27. $[\mathfrak{g}_\lambda, \mathfrak{g}_\mu] \subset \mathfrak{g}_{\lambda+\mu}$.

Proof. By Jacobi identity, $(\operatorname{ad} x - \lambda - \mu)[y, z] = [(\operatorname{ad} x - \lambda)y, z] + [y, (\operatorname{ad} x - \mu)z]$. Thus, if $y \in \mathfrak{g}_\lambda, z \in \mathfrak{g}_\mu$, then induction gives

$$(\operatorname{ad} x - \lambda - \mu)^n[y, z] = \sum_k \binom{n}{k}[(\operatorname{ad} x - \lambda)^k y, (\operatorname{ad} x - \mu)^{n-k}z],$$

which is zero for $n > \dim \mathfrak{g}_\lambda + \dim \mathfrak{g}_\mu$. $\qquad \square$

Let $\operatorname{ad} x = (\operatorname{ad} x)_s + (\operatorname{ad} x)_n$ be the Jordan decomposition of operator $\operatorname{ad} x$ (see Theorem 5.59), so that $(\operatorname{ad} x)_s|_{\mathfrak{g}_\lambda} = \lambda$. Then the lemma implies that $(\operatorname{ad} x)_s$ is a derivation of $\mathfrak{g}$. By Proposition 6.8, any derivation is inner, so $(\operatorname{ad} x)_s = \operatorname{ad} x_s$ for some $x_s \in \mathfrak{g}$; thus, $(\operatorname{ad} x)_n = \operatorname{ad}(x - x_s)$. This proves the existence of the Jordan decomposition for x. It also shows that if $\operatorname{ad} x.y = 0$, then $(\operatorname{ad} x)_s.y = (\operatorname{ad} x_s).y = 0$. $\qquad \square$

Corollary 6.28. *In any semisimple complex Lie algebra, there exist non-zero semisimple elements.*

Proof. If any semisimple element is zero, then, by Theorem 6.26, any $x \in \mathfrak{g}$ is nilpotent. By Engel's theorem (Theorem 5.34), this implies that $\mathfrak{g}$ is nilpotent and thus solvable, which contradicts the semisimplicity of $\mathfrak{g}$. $\qquad \square$

Our next step would be considering not just one semisimple element but a family of commuting semisimple elements.

Definition 6.29. A subalgebra $\mathfrak{h} \subset \mathfrak{g}$ is called *toral* if it is commutative and consists of semisimple elements.

Theorem 6.30. *Let $\mathfrak{g}$ be a complex semisimple Lie algebra, $\mathfrak{h} \subset \mathfrak{g}$ a toral subalgebra, and let (,) be a non-degenerate invariant symmetric bilinear form on $\mathfrak{g}$ (for example, the Killing form). Then*

(1) $\mathfrak{g} = \bigoplus_{\alpha \in \mathfrak{h}^*} \mathfrak{g}_\alpha$, *where $\mathfrak{g}_\alpha$ is a common eigenspace for all operators* $\operatorname{ad} h, h \in \mathfrak{h}$, *with eigenvalue α:*

$$\operatorname{ad} h.x = \langle \alpha, h \rangle x, \quad h \in \mathfrak{h}, x \in \mathfrak{g}_\alpha.$$

In particular, $\mathfrak{h} \subset \mathfrak{g}_0$.

(2) $[\mathfrak{g}_\alpha, \mathfrak{g}_\beta] \subset \mathfrak{g}_{\alpha+\beta}$.
(3) *If $\alpha + \beta \neq 0$, then $\mathfrak{g}_\alpha, \mathfrak{g}_\beta$ are orthogonal with respect to the form (,).*
(4) *For any α, the form (,) gives a non-degenerate pairing $\mathfrak{g}_\alpha \otimes \mathfrak{g}_{-\alpha} \to \mathbb{C}$.*

Proof. By definition, for each $h \in \mathfrak{h}$, the operator $\operatorname{ad} h$ is diagonalizable. Since all operators $\operatorname{ad} h$ commute, they can be simultaneously diagonalized, which is exactly the statement of the first part of the theorem. Of course, since $\mathfrak{g}$ is finite-dimensional, $\mathfrak{g}_\alpha = 0$ for all but finitely many $\alpha \subset \mathfrak{h}^*$.

The second part is in fact a very special case of Lemma 6.27. However, in this case it can be proved much easier: if $y \in \mathfrak{g}_\alpha, z \in \mathfrak{g}_\beta$, then

$$\operatorname{ad} h.[y, z] = [\operatorname{ad} h.y, z] + [y, \operatorname{ad} h.z] = \langle \alpha, h \rangle [y, z]$$
$$+ \langle \beta, h \rangle [y, z] = \langle \alpha + \beta, h \rangle [y, z].$$

For the third part, notice that if $x \in \mathfrak{g}_\alpha, y \in \mathfrak{g}_\beta, h \in \mathfrak{h}$, then invariance of the form shows that $([h, x], y) + (x, [h, y]) = (\langle h, \alpha \rangle + \langle h, \beta \rangle)(x, y) = 0$; thus, if $(x, y) \neq 0$, then $\langle h, \alpha + \beta \rangle = 0$ for all h, which implies $\alpha + \beta = 0$.

The final part immediately follows from the previous part. $\qquad \square$

For future use, we will also need some information about the zero eigenvalue subspace $\mathfrak{g}_0$.

Lemma 6.31. *In the notation of Theorem 6.30, we have*

(1) *Restriction of the form (,) to $\mathfrak{g}_0$ is non-degenerate.*
(2) *Let $x \in \mathfrak{g}_0$ and let $x = x_s + x_n$ be the Jordan decomposition of x (see Theorem 6.26). Then $x_s, x_n \in \mathfrak{g}_0$.*
(3) *$\mathfrak{g}_0$ is a reductive subalgebra in $\mathfrak{g}$.*

Proof. Part (1) is a special case of the last part in Theorem 6.30.

To prove part (2), note that if $x \in \mathfrak{g}_0$, then $[x, h] = 0$ for all $h \in \mathfrak{h}$. But then, by Theorem 6.26, $[x_s, h] = 0$, so $x_s \in \mathfrak{g}_0$.

To prove the last part, consider $\mathfrak{g}$ as a representation of $\mathfrak{g}_0$. Then the trace form $(x_1, x_2) = \mathrm{tr}_{\mathfrak{g}}(\mathrm{ad}\, x_1 \, \mathrm{ad}\, x_2)$ on $\mathfrak{g}_0$ is exactly the restriction of the Killing form $K^{\mathfrak{g}}$ to $\mathfrak{g}_0$ and by part (1) is non-degenerate. But by one of the forms of Cartan's criterion (Theorem 5.48), this implies that $\mathfrak{g}_0$ is reductive. $\qquad\square$

6.5. Cartan subalgebra

Our next goal is to produce as large a toral subalgebra in $\mathfrak{g}$ as possible. The standard way of formalizing this is as follows.

Definition 6.32. Let $\mathfrak{g}$ be a complex semisimple Lie algebra. A Cartan subalgebra $\mathfrak{h} \subset \mathfrak{g}$ is a toral subalgebra which coincides with its centralizer: $C(\mathfrak{h}) = \{x \mid [x, \mathfrak{h}] = 0\} = \mathfrak{h}$.

Remark 6.33. This definition should only be used for semisimple Lie algebras: for general Lie algebras, Cartan subalgebras are defined in a different way (see, e.g., [47]). However, it can be shown that for semisimple algebras our definition is equivalent to the usual one. (Proof in one direction is given in Exercise 6.3.)

Example 6.34. Let $\mathfrak{g} = \mathfrak{sl}(n, \mathbb{C})$ and $\mathfrak{h} = \{$diagonal matrices with trace $0\}$. Then $\mathfrak{h}$ is a Cartan subalgebra. Indeed, it is obviously commutative, and every diagonal element is semisimple (see Exercise 6.2), so it is a toral subalgebra. On the other hand, choose $h \in \mathfrak{h}$ to be a diagonal matrix with distinct eigenvalues. By a well-known result of linear algebra, if $[x, h] = 0$, and h has distinct eigenvalues, then any eigenvector of h is also an eigenvector of x; thus, x must also be diagonal. Thus, $C(h) = \mathfrak{h}$.

We still need to prove existence of Cartan subalgebras.

Theorem 6.35. *Let* $\mathfrak{h} \subset \mathfrak{g}$ *be a maximal toral subalgebra, i.e. a toral subalgebra which is not properly contained in any other toral subalgebra. Then* $\mathfrak{h}$ *is a Cartan subalgebra.*

Proof. Let $\mathfrak{g} = \bigoplus \mathfrak{g}_\alpha$ be the decomposition of $\mathfrak{g}$ into eigenspaces for $\mathrm{ad}\, h$ as in Theorem 6.30. We will show that $\mathfrak{g}_0 = C(\mathfrak{h})$ is toral; since it contains $\mathfrak{h}$, this would imply that $C(\mathfrak{h}) = \mathfrak{h}$, as desired.

First, note that for any $x \in \mathfrak{g}_0$, operator $\mathrm{ad}\, x|_{\mathfrak{g}_0}$ is nilpotent: otherwise, $\mathrm{ad}\, x|_{\mathfrak{g}_0}$ would have non-zero eigenvalues. Then the semisimple part x_s (which, by Lemma 6.31, is also in $\mathfrak{g}_0$) would be a semisimple element satisfying

$(\operatorname{ad} x_s)|_{\mathfrak{g}_0} \neq 0$ and thus $x_s \notin \mathfrak{h}$. On the other hand, $[\mathfrak{h}, x_s] = 0$ (since $x_s \in \mathfrak{g}_0$), so $\mathfrak{h} \oplus \mathbb{C} \cdot x_s$ would be a toral subalgebra, which contradicts maximality of $\mathfrak{h}$.

By Engel's theorem (Theorem 5.34), this implies that $\mathfrak{g}_0$ is nilpotent. On the other hand, by Lemma 6.31, $\mathfrak{g}_0$ is reductive. Therefore, it must be commutative.

Finally, to show that any $x \in \mathfrak{g}_0$ is semisimple, it suffices to show that for any such x, the nilpotent part $x_n = 0$ (recall that by Lemma 6.31, $x_n \in \mathfrak{g}_0$). But since $\operatorname{ad} x_n$ is nilpotent and $\mathfrak{g}_0$ is commutative, for any $y \in \mathfrak{g}_0$, $\operatorname{ad} x_n \operatorname{ad} y$ is also nilpotent, so $\operatorname{tr}_{\mathfrak{g}}(\operatorname{ad} x_n \operatorname{ad} y) = 0$. Since the Killing form on $\mathfrak{g}_0$ is non-degenerate (Lemma 6.31), this implies $x_n = 0$. Thus, we see that $\mathfrak{g}_0 = C(\mathfrak{h})$ is a toral subalgebra which contains $\mathfrak{h}$. Since $\mathfrak{h}$ was chosen to be maximal, $C(\mathfrak{h}) = \mathfrak{h}$. $\square$

Corollary 6.36. *In every complex semisimple Lie algebra $\mathfrak{g}$, there exists a Cartan subalgebra.*

Later (see Section 6.7) we will give another way of constructing Cartan subalgebras and will prove that all Cartan subalgebras are actually conjugate in $\mathfrak{g}$. In particular, this implies that they have the same dimension. This dimension is called the *rank* of $\mathfrak{g}$:

$$\operatorname{rank}(\mathfrak{g}) = \dim \mathfrak{h}. \tag{6.1}$$

Example 6.37. Rank of $\mathfrak{sl}(n, \mathbb{C})$ is equal to $n - 1$.

6.6. Root decomposition and root systems

From now on, we fix a complex semisimple Lie algebra $\mathfrak{g}$ and a Cartan subalgebra $\mathfrak{h} \subset \mathfrak{g}$.

Theorem 6.38.

(1) *We have the following decomposition for $\mathfrak{g}$, called the* root decomposition

$$\mathfrak{g} = \mathfrak{h} \oplus \bigoplus_{\alpha \in R} \mathfrak{g}_\alpha, \tag{6.2}$$

where

$$\mathfrak{g}_\alpha = \{x \mid [h, x] = \langle \alpha, h \rangle x \text{ for all } h \in \mathfrak{h}\}$$
$$R = \{\alpha \in \mathfrak{h}^* - \{0\} \mid \mathfrak{g}_\alpha \neq 0\}. \tag{6.3}$$

The set R is called the root system *of $\mathfrak{g}$, and subspaces $\mathfrak{g}_\alpha$ are called the* root subspaces.

(2) $[\mathfrak{g}_\alpha, \mathfrak{g}_\beta] \subset \mathfrak{g}_{\alpha+\beta}$ *(here and below, we let $\mathfrak{g}_0 = \mathfrak{h}$).*

(3) *If* $\alpha + \beta \neq 0$, *then* $\mathfrak{g}_\alpha, \mathfrak{g}_\beta$ *are orthogonal with respect to the Killing form K.*

(4) *For any* α, *the Killing form gives a non-degenerate pairing* $\mathfrak{g}_\alpha \otimes \mathfrak{g}_{-\alpha} \to \mathbb{C}$. *In particular, restriction of K to* $\mathfrak{h}$ *is non-degenerate.*

Proof. This immediately follows from Theorem 6.30 and $\mathfrak{g}_0 = \mathfrak{h}$, which is the definition of Cartan subalgebra. $\square$

Theorem 6.39. *Let* $\mathfrak{g}_1 \ldots \mathfrak{g}_n$ *be simple Lie algebras and let* $\mathfrak{g} = \bigoplus \mathfrak{g}_i$.

(1) *Let* $\mathfrak{h}_i \subset \mathfrak{g}_i$ *be Cartan subalgebras of* $\mathfrak{g}_i$ *and* $R_i \subset \mathfrak{h}_i^*$ *the corresponding root systems of* $\mathfrak{g}_i$. *Then* $\mathfrak{h} = \bigoplus \mathfrak{h}_i$ *is a Cartan subalgebra in* $\mathfrak{g}$ *and the corresponding root system is* $R = \sqcup R_i$.

(2) *Each Cartan subalgebra in* $\mathfrak{g}$ *must have the form* $\mathfrak{h} = \bigoplus \mathfrak{h}_i$ *where* $\mathfrak{h}_i \subset \mathfrak{g}_i$ *is a Cartan subalgebra in* $\mathfrak{g}_i$.

Proof. The first part is obvious from the definitions. To prove the second part, let $\mathfrak{h}_i = \pi_i(\mathfrak{h})$, where $\pi_i : \mathfrak{g} \to \mathfrak{g}_i$ is the projection. It is immediate that for $x \in \mathfrak{g}_i, h \in \mathfrak{h}$, we have $[h, x] = [\pi_i(h), x]$. From this it easily follows that $\mathfrak{h}_i$ is a Cartan subalgebra. To show that $\mathfrak{h} = \bigoplus \mathfrak{h}_i$, notice that obviously $\mathfrak{h} \subset \bigoplus \mathfrak{h}_i$; since $\bigoplus \mathfrak{h}_i$ is toral, by definition of Cartan subalgebra we must have $\mathfrak{h} = \bigoplus \mathfrak{h}_i$. $\square$

Example 6.40. Let $\mathfrak{g} = \mathfrak{sl}(n, \mathbb{C}), \mathfrak{h} = $ diagonal matrices with trace 0 (see Example 6.34). Denote by $e_i : \mathfrak{h} \to \mathbb{C}$ the functional which computes i^{th} diagonal entry of h:

$$
e_i : \begin{bmatrix} h_1 & 0 & \cdots \\ 0 & h_2 & \cdots \\ & \cdots & \\ 0 & \cdots & h_n \end{bmatrix} \mapsto h_i.
$$

Then one easily sees that $\sum e_i = 0$, so

$$
\mathfrak{h}^* = \bigoplus \mathbb{C}e_i / \mathbb{C}(e_1 + \cdots + e_n).
$$

It is easy to see that matrix units E_{ij} are eigenvectors for ad $h, h \in \mathfrak{h}$: $[h, E_{ij}] = (h_i - h_j)E_{ij} = (e_i - e_j)(h)E_{ij}$. Thus, the root decomposition is given by

$$
R = \{e_i - e_j \mid i \neq j\} \subset \bigoplus \mathbb{C}e_i / \mathbb{C}(e_1 + \cdots + e_n)
$$

$$
\mathfrak{g}_{e_i - e_j} = \mathbb{C}E_{ij}.
$$

The Killing form on $\mathfrak{h}$ is given by

$$(h, h') = \sum_{i \neq j}(h_i - h_j)(h'_i - h'_j) = 2n \sum_i h_i h'_i = 2n \operatorname{tr}(hh').$$

From this, it is easy to show that if $\lambda = \sum \lambda_i e_i, \mu = \sum \mu_i e_i \in \mathfrak{h}^*$, and λ_i, μ_i are chosen so that $\sum \lambda_i = \sum \mu_i = 0$ (which is always possible), then the corresponding form on $\mathfrak{h}^*$ is given by

$$(\lambda, \mu) = \frac{1}{2n} \sum_i \lambda_i \mu_i.$$

The root decomposition is the most important result one should know about semisimple Lie algebras – much more important than the definition of semisimple algebras (in fact, this could be taken as the definition, see Exercise 6.4). Our goal is to use this decomposition to get as much information as possible about the structure of semisimple Lie algebras, eventually getting full classification theorem for them.

From now on, we will denote by (,) a non-degenerate symmetric invariant bilinear form on $\mathfrak{g}$. Such a form exists: for example, one can take (,) to be the Killing form (in fact, if $\mathfrak{g}$ is simple, then any invariant bilinear form is a multiple of the Killing form, see Exercise 4.5). However, in most cases it is more convenient to use a different normalization, which we will introduce later, in Exercise 8.7.

Since the restriction of (,) to $\mathfrak{h}$ is non-degenerate (see Theorem 6.38), it defines an isomorphism $\mathfrak{h} \xrightarrow{\sim} \mathfrak{h}^*$ and a non-degenerate bilinear form on $\mathfrak{h}^*$, which we will also denote by (,). It can be explicitly defined as follows: if we denote for $\alpha \in \mathfrak{h}^*$ by H_α the corresponding element of $\mathfrak{h}$, then

$$(\alpha, \beta) = \langle H_\alpha, \beta \rangle = (H_\alpha, H_\beta) \tag{6.4}$$

for any $\alpha, \beta \in \mathfrak{h}^*$.

Lemma 6.41. *Let $e \in \mathfrak{g}_\alpha, f \in \mathfrak{g}_{-\alpha}$ and let H_α be defined by (6.4). Then*

$$[e, f] = (e, f)H_\alpha.$$

Proof. Let us compute the pairing $([e, f], h)$ for some $h \in \mathfrak{h}$. Since the form (,) is invariant, we have

$$([e, f], h) = (e, [f, h]) = -(e, [h, f]) = \langle h, \alpha \rangle(e, f) = (e, f)(h, H_\alpha)$$

Since (,) is a non-degenerate form on $\mathfrak{h}$, this implies that $[e, f] = (e, f)H_\alpha$. $\quad\square$

Lemma 6.42.

(1) *Let $\alpha \in R$. Then $(\alpha, \alpha) = (H_\alpha, H_\alpha) \neq 0$.*
(2) *Let $e \in \mathfrak{g}_\alpha, f \in \mathfrak{g}_{-\alpha}$ be such that $(e,f) = 2/(\alpha, \alpha)$, and let*

$$h_\alpha = \frac{2H_\alpha}{(\alpha, \alpha)}. \tag{6.5}$$

Then $\langle h_\alpha, \alpha \rangle = 2$ and the elements e, f, h_α satisfy the commutation relations (3.24) of Lie algebra $\mathfrak{sl}(2, \mathbb{C})$. We will denote such a subalgebra by $\mathfrak{sl}(2, \mathbb{C})_\alpha \subset \mathfrak{g}$.
(3) *So defined h_α is independent of the choice of non-degenerate invariant bilinear form $(,)$.*

Proof. Assume that $(\alpha, \alpha) = 0$; then $\langle H_\alpha, \alpha \rangle = 0$. Choose $e \in \mathfrak{g}_\alpha, f \in \mathfrak{g}_{-\alpha}$ such that $(e,f) \neq 0$ (possible by Theorem 6.38). Let $h = [e,f] = (e,f)H_\alpha$ and consider the algebra $\mathfrak{a}$ generated by e, f, h. Then we see that $[h, e] = \langle h, \alpha \rangle e = 0, [h, f] = -\langle h, \alpha \rangle f = 0$, so $\mathfrak{a}$ is solvable Lie algebra. By Lie theorem (Theorem 5.30), we can choose a basis in $\mathfrak{g}$ such that operators ad e, ad f, ad h are upper triangular. Since $h = [e,f]$, ad h will be strictly upper-triangular and thus nilpotent. But since $h \in \mathfrak{h}$, it is also semisimple. Thus, $h = 0$. On the other hand, $h = (e,f)H_\alpha \neq 0$. This contradiction proves the first part of the theorem.

The second part is immediate from definitions and Lemma 6.41.

The last part is left as an exercise to the reader (Exercise 6.6). □

This lemma gives us a very powerful tool for study of $\mathfrak{g}$: we can consider $\mathfrak{g}$ as a module over the subalgebra $\mathfrak{sl}(2, \mathbb{C})_\alpha$ and then use results about representations of $\mathfrak{sl}(2, \mathbb{C})$ proved in Section 4.8.

Lemma 6.43. *Let α be a root, and let $\mathfrak{sl}(2, \mathbb{C})_\alpha$ be the Lie subalgebra generated by $e \in \mathfrak{g}_\alpha, f \in \mathfrak{g}_{-\alpha}$ and h_α as in Lemma 6.42.*

Consider the subspace

$$V = \mathbb{C}h_\alpha \oplus \bigoplus_{k \in \mathbb{Z}, k \neq 0} \mathfrak{g}_{k\alpha} \subset \mathfrak{g}.$$

Then V is an irreducible representation of $\mathfrak{sl}(2, \mathbb{C})_\alpha$.

Proof. Since ad $e.\mathfrak{g}_{k\alpha} \subset \mathfrak{g}_{(k+1)\alpha}$, and by Lemma 6.41, ad $e.\mathfrak{g}_{-\alpha} \subset \mathbb{C}h_\alpha$, and similarly for f, V is a representation of $\mathfrak{sl}(2, \mathbb{C})_\alpha$. Since $\langle h_\alpha, \alpha \rangle = 2$, we see that the weight decomposition of V is given by $V[k] = 0$ for odd k and $V[2k] = \mathfrak{g}_{k\alpha}$, $V[0] = \mathbb{C}h_\alpha$. In particular, zero weight space $V[0]$ is one-dimensional. By Exercise 4.11, this implies that V is irreducible. □

Now we can prove the main theorem about the structure of semisimple Lie algebras.

Theorem 6.44. *Let $\mathfrak{g}$ be a complex semisimple Lie algebra with Cartan subalgebra $\mathfrak{h}$ and root decomposition $\mathfrak{g} = \mathfrak{h} \oplus \bigoplus_{\alpha \in R} \mathfrak{g}_\alpha$. Let $(\,,)$ a non-degenerate symmetric invariant bilinear form on $\mathfrak{g}$.*

(1) *R spans $\mathfrak{h}^*$ as a vector space, and elements $h_\alpha, \alpha \in R$, defined by (6.5) span $\mathfrak{h}$ as a vector space.*
(2) *For each $\alpha \in R$, the root subspace $\mathfrak{g}_\alpha$ is one-dimensional.*
(3) *For any two roots α, β, the number*

$$\langle h_\alpha, \beta \rangle = \frac{2(\alpha, \beta)}{(\alpha, \alpha)}$$

is integer.
(4) *For $\alpha \in R$, define the reflection operator $s_\alpha : \mathfrak{h}^* \to \mathfrak{h}^*$ by*

$$s_\alpha(\lambda) = \lambda - \langle h_\alpha, \lambda \rangle \alpha = \lambda - \frac{2(\alpha, \lambda)}{(\alpha, \alpha)} \alpha.$$

Then for any roots $\alpha, \beta, s_\alpha(\beta)$ is also a root. In particular, if $\alpha \in R$, then $-\alpha = s_\alpha(\alpha) \in R$.
(5) *For any root α, the only multiples of α which are also roots are $\pm\alpha$.*
(6) *For roots $\alpha, \beta \neq \pm\alpha$, the subspace*

$$V = \bigoplus_{k \in \mathbb{Z}} \mathfrak{g}_{\beta+k\alpha}$$

is an irreducible representation of $\mathfrak{sl}(2, \mathbb{C})_\alpha$.
(7) *If α, β are roots such that $\alpha + \beta$ is also a root, then $[\mathfrak{g}_\alpha, \mathfrak{g}_\beta] = \mathfrak{g}_{\alpha+\beta}$.*

Proof. (1) Assume that R does not generate $\mathfrak{h}^*$; then there exists a non-zero $h \in \mathfrak{h}$ such that $\langle h, \alpha \rangle = 0$ for all $\alpha \in R$. But then root decomposition (6.2) implies that $\mathrm{ad}\, h = 0$. However, by definition in a semisimple Lie algebra, the center is trivial: $\mathfrak{z}(\mathfrak{g}) = 0$.

The fact that h_α span $\mathfrak{h}$ now immediately follows: using identification of $\mathfrak{h}$ with $\mathfrak{h}^*$ given by the Killing form, elements h_α are identified with non-zero multiples of α.

(2) Immediate from Lemma 6.43 and the fact that in any irreducible representation of $\mathfrak{sl}(2, \mathbb{C})$, weight subspaces are one-dimensional.

(3) Consider $\mathfrak{g}$ as a representation of $\mathfrak{sl}(2,\mathbb{C})_\alpha$. Then elements of $\mathfrak{g}_\beta$ have weight equal to $\langle h_\alpha, \beta \rangle$. But by Theorem 4.60, weights of any finite-dimensional representation of $\mathfrak{sl}(2,\mathbb{C})$ are integer.

(4) Assume that $\langle h_\alpha, \beta \rangle = n \geq 0$. Then elements of $\mathfrak{g}_\beta$ have weight n with respect to the action of $\mathfrak{sl}(2,\mathbb{C})_\alpha$. By Theorem 4.60, operator f_α^n is an isomorphism of the space of vectors of weight n with the space of vectors of weight $-n$. In particular, it means that if $v \in \mathfrak{g}_\beta$ is non-zero vector, then $f_\alpha^n v \in \mathfrak{g}_{\beta-n\alpha}$ is also non-zero. Thus, $\beta - n\alpha = s_\alpha(\beta) \in R$.

For $n \leq 0$, the proof is similar, using e^{-n} instead of f^n.

(5) Assume that α and $\beta = c\alpha, c \in \mathbb{C}$, are both roots. By part (3), $(2(\alpha, \beta)/(\alpha, \alpha)) = 2c$ is integer, so c is a half-integer. The same argument shows that $1/c$ is also a half-integer. It is easy to see that this implies that $c \in \{\pm 1, \pm 2, \pm 1/2\}$. Interchanging the roots if necessary and possibly replacing α by $-\alpha$, we have $c = 1$ or $c = 2$.

Now let us consider the subspace

$$V = \mathbb{C}h_\alpha \oplus \bigoplus_{k \in \mathbb{Z}, k \neq 0} \mathfrak{g}_{k\alpha} \subset \mathfrak{g}.$$

By Lemma 6.43, V is an irreducible representation of $\mathfrak{sl}(2,\mathbb{C})_\alpha$, and by part (2), $V[2] = \mathfrak{g}_\alpha = \mathbb{C}e_\alpha$. Thus, the map ad $e_\alpha : \mathfrak{g}_\alpha \to \mathfrak{g}_{2\alpha}$ is zero. But the results of Section 4.8 show that in an irreducible representation, the kernel of e is exactly the highest weight subspace. Thus, we see that V has highest weight 2: $V[4] = V[6] = \cdots = 0$. This means that $V = \mathfrak{g}_{-\alpha} \oplus \mathbb{C}h_\alpha \oplus \mathfrak{g}_\alpha$, so the only integer multiples of α which are roots are $\pm\alpha$. In particular, 2α is not a root.

Combining these two results, we see that if $\alpha, c\alpha$ are both roots, then $c = \pm 1$.

(6) Proof is immediate from dim $\mathfrak{g}_{\beta+k\alpha} = 1$.

(7) We already know that $[\mathfrak{g}_\alpha, \mathfrak{g}_\beta] \subset \mathfrak{g}_{\alpha+\beta}$. Since dim $\mathfrak{g}_{\alpha+\beta} = 1$, we need to show that for non-zero $e_\alpha \in \mathfrak{g}_\alpha, e_\beta \in \mathfrak{g}_\beta$, we have $[e_\alpha, e_\beta] \neq 0$. This follows from the previous part and the fact that in an irreducible representation of $\mathfrak{sl}(2,\mathbb{C})$, if $v \in V[k]$ is non-zero and $V[k+2] \neq 0$, then $e.v \neq 0$. $\square$

In the next chapter, we will study the set of roots R in detail. As we will see, it gives us a key to the classification of semisimple Lie algebras.

Theorem 6.45.

(1) *Let* $\mathfrak{h}_\mathbb{R} \subset \mathfrak{h}$ *be the real vector space generated by* $h_\alpha, \alpha \in R$. *Then* $\mathfrak{h} = \mathfrak{h}_\mathbb{R} \oplus i\mathfrak{h}_\mathbb{R}$, *and the restriction of the Killing form to* $\mathfrak{h}_\mathbb{R}$ *is positive definite.*

(2) *Let* $\mathfrak{h}_{\mathbb{R}}^* \subset \mathfrak{h}^*$ *be the real vector space generated by* $\alpha \in R$. *Then* $\mathfrak{h}^* = \mathfrak{h}_{\mathbb{R}}^* \oplus i\mathfrak{h}_{\mathbb{R}}^*$. *Also,* $\mathfrak{h}_{\mathbb{R}}^* = \{\lambda \in \mathfrak{h}^* \mid \langle \lambda, h \rangle \in \mathbb{R} \text{ for all } h \in \mathfrak{h}_{\mathbb{R}}\} = (\mathfrak{h}_{\mathbb{R}})^*$.

Proof. Let us first prove that the restriction of the Killing form to $\mathfrak{h}_{\mathbb{R}}$ is real and positive definite. Indeed,

$$(h_\alpha, h_\beta) = \operatorname{tr}(\operatorname{ad} h_\alpha \operatorname{ad} h_\beta) = \sum_{\gamma \in R} \langle h_\alpha, \gamma \rangle \langle h_\beta, \gamma \rangle.$$

But by Theorem 6.44, $\langle h_\alpha, \gamma \rangle, \langle h_\beta, \gamma \rangle \in \mathbb{Z}$, so $(h_\alpha, h_\beta) \in \mathbb{Z}$.

Now let $h = \sum c_\alpha h_\alpha \in \mathfrak{h}_{\mathbb{R}}$. Then $\langle h, \gamma \rangle = \sum c_\alpha \langle h_\alpha, \gamma \rangle \in \mathbb{R}$ for any root γ, so

$$(h, h) = \operatorname{tr}(\operatorname{ad} h)^2 = \sum_\gamma \langle h, \gamma \rangle^2 \geq 0$$

which proves that the Killing form is positive definite on $\mathfrak{h}_{\mathbb{R}}$.

Since the Killing form is positive definite on $\mathfrak{h}_{\mathbb{R}}$, it is negative definite on $i\mathfrak{h}_{\mathbb{R}}$, so $\mathfrak{h}_{\mathbb{R}} \cap i\mathfrak{h}_{\mathbb{R}} = \{0\}$, which implies $\dim_{\mathbb{R}} \mathfrak{h}_{\mathbb{R}} \leq \frac{1}{2} \dim_{\mathbb{R}} \mathfrak{h} = r$, where $r = \dim_{\mathbb{C}} \mathfrak{h}$ is the rank of $\mathfrak{g}$. On the other hand, since h_α generate $\mathfrak{h}$ over $\mathbb{C}$, we see that $\dim_{\mathbb{R}} \mathfrak{h}_{\mathbb{R}} \geq r$. Thus, $\dim_{\mathbb{R}} \mathfrak{h}_{\mathbb{R}} = r$, so $\mathfrak{h} = \mathfrak{h}_{\mathbb{R}} \oplus i\mathfrak{h}_{\mathbb{R}}$.

The second part easily follows from the first one. $\qquad\square$

Remark 6.46. It easily follows from this theorem and Theorem 6.10 that if $\mathfrak{k}$ is a compact real form of $\mathfrak{g}$ (see Theorem 6.13), then $\mathfrak{k} \cap \mathfrak{h} = i\mathfrak{h}_{\mathbb{R}}$. For example, for $\mathfrak{g} = \mathfrak{sl}(n, \mathbb{C})$, $\mathfrak{h}_{\mathbb{R}}$ consists of traceless diagonal matrices with real entries, and $\mathfrak{su}(n) \cap \mathfrak{h} =$ traceless diagonal skew-hermitian matrices $= i\mathfrak{h}_{\mathbb{R}}$.

6.7. Regular elements and conjugacy of Cartan subalgebras

In this section, we give another way of constructing Cartan subalgebras, and prove conjugacy of Cartan subalgebras, which was stated without proof in Section 6.5. This section can be skipped at first reading.

We start with an example.

Example 6.47. Let $\mathfrak{g} = \mathfrak{sl}(n, \mathbb{C})$ and let $h \in \mathfrak{g}$ be such that all eigenvalues of h are distinct (as is well known, set of such matrices is open and dense in $\mathfrak{g}$). In this case, h has an eigenbasis v_i in which it becomes a diagonal matrix with distinct numbers λ_i on the diagonal. Since eigenvalues of $\operatorname{ad} h$ are of the form $\lambda_i - \lambda_j$ (see proof of Theorem 5.60), this implies that $[h, x] = 0$ iff x itself is diagonal in the basis v_i. Therefore, in this case the subalgebra of diagonal matrices (which, as discussed in Example 6.34, is a Cartan subalgebra in $\mathfrak{sl}(n, \mathbb{C})$) can be recovered as the centralizer of h: $\mathfrak{h} = C(h)$.

This suggests a method of constructing Cartan subalgebras for an arbitrary semisimple Lie algebra $\mathfrak{g}$ as centralizers of the "generic" element $h \in \mathfrak{g}$. We must, however, give a precise definition of the word "generic", which can be done as follows.

Definition 6.48. For any $x \in \mathfrak{g}$, define "nullity" of x by

$$n(x) = \text{multiplicity of } 0 \text{ as a generalized eigenvalue of } \text{ad } x$$

It is clear that for every $x \in \mathfrak{g}, n(x) \geq 1$ (because x itself is annihilated by $\text{ad } x$).

Definition 6.49. For any Lie algebra $\mathfrak{g}$, its rank $\text{rank}(\mathfrak{g})$ is defined by

$$\text{rank}(\mathfrak{g}) = \min_{x \in \mathfrak{g}} n(x)$$

An element $x \in \mathfrak{g}$ is called *regular* if $n(x) = \text{rank}(\mathfrak{g})$.

Example 6.50. Let $\mathfrak{g} = \mathfrak{gl}(n, \mathbb{C})$ and let $x \in \mathfrak{g}$ have eigenvalues λ_i. Then eigenvalues of $\text{ad } x$ are $\lambda_i - \lambda_j, i, j = 1 \ldots n$; thus, $n(x) \geq n$ with the equality iff all λ_i are distinct. Therefore, $\text{rank}(\mathfrak{gl}(n, \mathbb{C})) = n$ and $x \in \mathfrak{gl}(n, \mathbb{C})$ is regular iff its eigenvalues are distinct (in which case it must be diagonalizable). A minor modification of this argument shows that $\text{rank}(\mathfrak{sl}(n, \mathbb{C})) = n - 1$ and $x \in \mathfrak{sl}(n, \mathbb{C})$ is regular iff its eigenvalues are distinct.

Lemma 6.51. *In any finite-dimensional complex Lie algebra* $\mathfrak{g}$ *, the set* $\mathfrak{g}^{reg}$ *of regular elements is connected, open and dense in* $\mathfrak{g}$.

Proof. For any $x \in \mathfrak{g}$, let $p_x(t) = \det(\text{ad } x - t) = a_n(x)t^n + \cdots + a_0(x)$ be the characteristic polynomial of $\text{ad } x$. By definition, $n(x)$ is multiplicity of zero as a root of $p_x(t)$: thus, for any x, polynomial $p_x(t)$ has zero of order $\geq r$ at zero, with equality iff x is regular (r is the rank of $\mathfrak{g}$). Therefore, x is regular iff $a_r(x) \neq 0$. However, each of the coefficients $a_k(x)$ is a polynomial function on $\mathfrak{g}$; thus, the set $\{x \mid a_r(x) \neq 0\}$ is an open dense set in $\mathfrak{g}$.

To prove that $\mathfrak{g}^{reg}$ is connected, note that for any affine complex line $l = x + ty, t \in \mathbb{C}$, the intersection

$$l \cap \mathfrak{g}^{reg} = \{x + ty \mid a_r(x + ty) \neq 0\}$$

is either empty (if $a_r(x + ty)$ is identically zero) or a complement of a finite number of points in l; in particular, if non-empty, it must be connected. Thus, if $x_1, x_2 \in \mathfrak{g}^{reg}$, by taking a complex line through them we see that x_1, x_2 are path-connected.   □

Proposition 6.52. *Let $\mathfrak{g}$ be a complex semisimple Lie algebra, and $\mathfrak{h} \subset \mathfrak{g}$ – a Cartan subalgebra.*

(1) $\dim \mathfrak{h} = \operatorname{rank}(\mathfrak{g})$.
(2)

$$\mathfrak{h} \cap \mathfrak{g}^{reg} = \{h \in \mathfrak{h} \mid \langle h, \alpha \rangle \neq 0 \; \forall \, \alpha \in R\}.$$

In particular, $\mathfrak{h} \cap \mathfrak{g}^{reg}$ is open and dense in $\mathfrak{h}$.

Proof. Let G be a connected Lie group with Lie algebra $\mathfrak{g}$ and define

$$V = \{h \in \mathfrak{h} \mid \langle h, \alpha \rangle \neq 0 \; \forall \, \alpha \in R\} \subset \mathfrak{h}$$

and let $U = \operatorname{Ad} G.V$. Then the set U is open in $\mathfrak{g}$. Indeed, consider the map $\varphi \colon G \times V \to \mathfrak{g}$ given by $(g, x) \mapsto \operatorname{Ad} g.x$. Then, for any $x \in V$, the corresponding map of tangent spaces at $(1, x)$ is given by $\varphi_* \colon \mathfrak{g} \times \mathfrak{h} \to \mathfrak{g} : (y, h) \mapsto [y, x] + h$. Since $[\mathfrak{g}_\alpha, x] = \mathfrak{g}_\alpha$ (this is where we need that $\langle x, \alpha \rangle \neq 0 \; \forall \alpha$), we see that φ_* is surjective; therefore, the image of φ contains an open neighborhood of x. Since any $u \in U$ can be written in the form $u = \operatorname{Ad} g.x$ for some $x \in V$, this implies that for any $u \in U$, the set U contains an open neighborhood of u.

Since the set U is open, it must intersect with the open dense set $\mathfrak{g}^{reg}$. On the other hand, for any $u = \operatorname{Ad} g.x \in U$, we have $n(u) = n(x) = \dim C(x)$, where $C(x) = \{y \in \mathfrak{g} \mid [x, y] = 0\}$ is the centralizer of x, so $\operatorname{rank}(\mathfrak{g}) = n(x)$. But it easily follows from the root decomposition and the definition of V that for any $x \in V$, we have $C(x) = \mathfrak{h}$. Therefore, $\operatorname{rank}(\mathfrak{g}) = \dim \mathfrak{h}$.

The second part of the proposition now immediately follows from the observation that for any $h \in \mathfrak{h}$, we have

$$n(h) = \dim C(h) = \dim \mathfrak{h} + |\{\alpha \in R \mid \langle h, \alpha \rangle = 0\}|.   □$$

Theorem 6.53. *Let $\mathfrak{g}$ be a complex semisimple Lie algebra, and $x \in \mathfrak{g}$ – a regular semisimple element. Then the centralizer $C(x) = \{y \in \mathfrak{g} \mid [x, y] = 0\}$ is a Cartan subalgebra in $\mathfrak{g}$. Conversely, any Cartan subalgebra in $\mathfrak{g}$ can be written as a centralizer of a regular semisimple element.*

Proof. Let us decompose $\mathfrak{g}$ into direct sum of eigenspaces for $\operatorname{ad} x$: $\mathfrak{g} = \bigoplus \mathfrak{g}_\lambda, \lambda \in \mathbb{C}, \operatorname{ad} x|_{\mathfrak{g}_\lambda} = \lambda$. Then the results of Theorem 6.30, Lemma 6.31 apply; in particular, $\mathfrak{g}_0 = C(x)$ is a reductive subalgebra in $\mathfrak{g}$.

We claim that $C(x)$ is nilpotent. Indeed, by Engel's theorem (Theorem 5.34) it suffices to prove that for any $y \in C(x)$, the restriction of ad y to $C(x)$ is nilpotent. Consider element $x_t = x + ty \in C(x)$. Then for small values of t, we have ad $x_t|_{\mathfrak{g}/\mathfrak{g}_0}$ is invertible (since ad $x|_{\mathfrak{g}/\mathfrak{g}_0}$ is invertible), so the null space of x_t is contained in $\mathfrak{g}_0 = C(x)$. On the other hand, by definition of rank we have $n(x_t) \geq \operatorname{rank}(\mathfrak{g}) = \dim C(x)$ (the last equality holds because x is regular). Thus, ad $x_t|_{C(x)}$ is nilpotent for t close to zero; since x acts by zero in $C(x)$, this means that ad $y|_{C(x)}$ is nilpotent.

Now the same arguments as in the proof of Theorem 6.35 show that $\mathfrak{g}_0 = C(x)$ is a toral subalgebra. Since $x \in C(x)$, the centralizer of $C(x)$ is contained in $C(x)$; thus, $C(x)$ is a Cartan subalgebra.

The last part is obvious: if $\mathfrak{h}$ is a Cartan subalgebra, then by Proposition 6.52 it contains a regular semisimple element x and thus $\mathfrak{h} \subset C(x)$; since $\dim \mathfrak{h} = \operatorname{rank}(\mathfrak{g}) = \dim C(x)$, we see that $\mathfrak{h} = C(x)$. $\quad\square$

Corollary 6.54. *In a complex semisimple Lie algebra:*

(1) *Any regular element is semisimple.*
(2) *Any regular element is contained in a unique Cartan subalgebra.*

Proof. It is immediate from the definition that eigenvalues of ad x and ad $x_s = (\operatorname{ad} x)_s$ coincide; thus, if x is regular then so is x_s. Then by Theorem 6.53, $C(x_s)$ is a Cartan subalgebra. Since $x \in C(x_s)$, x itself must be semisimple.

To prove the second part, note that by (1) and Theorem 6.53, for any regular x the centralizer $C(x)$ is a Cartan subalgebra, so x is contained in a Cartan subalgebra. To prove uniqueness, note that if $\mathfrak{h} \ni x$ is a Cartan subalgebra, then commutativity of $\mathfrak{h}$ implies that $\mathfrak{h} \subset C(x)$. On the other hand, by Proposition 6.52, $\dim \mathfrak{h} = \operatorname{rank}(\mathfrak{g}) = \dim C(x)$. $\quad\square$

Theorem 6.55. *Any two Cartan subalgebras in a semisimple Lie algebra are conjugate: if $\mathfrak{h}_1, \mathfrak{h}_2 \subset \mathfrak{g}$ are Cartan subalgebras, then there exists an element g in the Lie group G corresponding to $\mathfrak{g}$ such that $\mathfrak{h}_2 = \operatorname{Ad} g(\mathfrak{h}_1)$.*

Proof. Consider the set $\mathfrak{g}^{\text{reg}}$ of regular elements in $\mathfrak{g}$; by Corollary 6.54, any such element is contained in a unique Cartan subalgebra, namely $\mathfrak{h}_x = C(x)$. Define the following equivalence relation on $\mathfrak{g}^{\text{reg}}$:

$$x \sim y \iff \mathfrak{h}_x \text{ is conjugate to } \mathfrak{h}_y$$

Obviously, if x, y are regular elements in the same Cartan subalgebra $\mathfrak{h}$, then for any $g \in G$, we have $y \sim \operatorname{Ad} g.x$. But it was shown in the proof of Proposition 6.52 that for fixed Cartan subalgebra $\mathfrak{h}$, the set $\{\operatorname{Ad} g.x \mid g \in G, x \in \mathfrak{h} \cap \mathfrak{g}^{\text{reg}}\}$

is open. Thus, each equivalence class of x contains a neighborhood of x and therefore is open.

Since the set $\mathfrak{g}^{\text{reg}}$ is connected (Lemma 6.51), and each equivalence class of relation $\sim$ is open, this implies that there is only one equivalence class: for any regular x, y, corresponding Cartan subalgebras $\mathfrak{h}_x, \mathfrak{h}_y$ are conjugate. Since every Cartan subalgebra has the form $\mathfrak{h}_x$ (Theorem 6.53), this implies the statement of the theorem. $\square$

6.8. Exercises

6.1. Show that the Casimir operator for $\mathfrak{g} = \mathfrak{so}(3, \mathbb{R})$ is given by $C = -1/2$ $(J_x^2 + J_y^2 + J_z^2)$, where generators J_x, J_y, J_z are defined in Section 3.10; thus, it follows from Proposition 6.15 that $J_x^2 + J_y^2 + J_z^2 \in U\mathfrak{so}(3, \mathbb{R})$ is central. Compare this with the proof of Lemma 4.62, where the same result was obtained by direct computation.

6.2. Show that for $\mathfrak{g} = \mathfrak{gl}(n, \mathbb{C})$, Definition 6.25 is equivalent to the usual definition of a semisimple operator (hint: use results of Section 5.9).

6.3. Show that if $\mathfrak{h} \subset \mathfrak{g}$ is a Cartan subalgebra in a complex semisimple Lie algebra, then $\mathfrak{h}$ is a nilpotent subalgebra which coincides with its normalizer $n(\mathfrak{h}) = \{x \in g \mid \text{ad} x.\mathfrak{h} \subset \mathfrak{h}\}$. (This is the usual definition of a Cartan subalgebra which can be used for any Lie algebra, not necessarily a semisimple one.)

6.4. Let $\mathfrak{h} \subset \mathfrak{so}(4, \mathbb{C})$ be the subalgebra consisting of matrices of the form

$$\begin{bmatrix} & a & & \\ -a & & & \\ & & & b \\ & & -b & \end{bmatrix}$$

(entries not shown are zeros). Show that then $\mathfrak{h}$ is a Cartan subalgebra and find the corresponding root decomposition.

6.5. (1) Define a bilinear form B on $W = \Lambda^2 \mathbb{C}^4$ by $\omega_1 \wedge \omega_2 = B(\omega_1, \omega_2) e_1 \wedge e_2 \wedge e_3 \wedge e_4$. Show that B is a symmetric non-degenerate form and construct an orthonormal basis for B.

(2) Let $\mathfrak{g} = \mathfrak{so}(W, B) = \{x \in \mathfrak{gl}(W) \mid B(x\omega_1, \omega_2) + B(\omega_1, x\omega_2) = 0\}$. Show that $\mathfrak{g} \simeq \mathfrak{so}(6, \mathbb{C})$.

(3) Show that the form B is invariant under the natural action of $\mathfrak{sl}(4,\mathbb{C})$ on $\Lambda^2\mathbb{C}^4$.

(4) Using results of the previous parts, construct a homomorphism $\mathfrak{sl}(4,\mathbb{C}) \to \mathfrak{so}(6,\mathbb{C})$ and prove that it is an isomorphism.

6.6. Show that definition (6.5) of h_α is independent of the choice of (,): replacing the Killing form by any other non-degenerate symmetric invariant bilinear form gives the same h_α. [Hint: show it first for a simple Lie algebra, then use Theorem 6.39.]

7

Root systems

7.1. Abstract root systems

The results of Section 6.6 show that the set of roots R of a semisimple complex Lie algebra has a number of remarkable properties. It turns out that sets with similar properties also appear in many other areas of mathematics. Thus, we will introduce the notion of abstract root system and study such objects, leaving for some time the theory of Lie algebras.

Definition 7.1. An abstract root system is a finite set of elements $R \subset E \setminus \{0\}$, where E is a Euclidean vector space (i.e., a real vector space with an inner product), such that the following properties hold:

(R1) R generates E as a vector space.
(R2) For any two roots α, β, the number

$$n_{\alpha\beta} = \frac{2(\alpha, \beta)}{(\beta, \beta)} \tag{7.1}$$

is integer.
(R3) Let $s_\alpha : E \to E$ be defined by

$$s_\alpha(\lambda) = \lambda - \frac{2(\alpha, \lambda)}{(\alpha, \alpha)} \alpha. \tag{7.2}$$

Then for any roots $\alpha, \beta, s_\alpha(\beta) \in R$.

The number $r = \dim E$ is called the *rank* of R.
  If, in addition, R satisfies the following property

(R4) If $\alpha, c\alpha$ are both roots, then $c = \pm 1$.

then R is called a *reduced* root system.

Remark 7.2. It is easy to deduce from (R1)–(R3) that if $\alpha, c\alpha$ are both roots, then $c \in \{\pm 1, \pm 2, \pm\frac{1}{2}\}$ (see the proof of Theorem 6.44). However, there are indeed examples of non-reduced root systems, which contain α and 2α as roots–see Exercise 7.1. Thus, condition (R4) does not follow from (R1)–(R3). However, in this book we will only consider reduced root systems.

Note that conditions (R2), (R3) have a very simple geometric meaning. Namely, s_α is the reflection around the hyperplane

$$L_\alpha = \{\lambda \in E \mid (\alpha, \lambda) = 0\}. \tag{7.3}$$

It can be defined by $s_\alpha(\lambda) = \lambda$ if $(\alpha, \lambda) = 0$ and $s_\alpha(\alpha) = -\alpha$.

Similarly, the number $n_{\alpha\beta}$ also has a simple geometric meaning: if we denote by p_α the operator of orthogonal projection onto the line containing α, then $p_\alpha(\beta) = (n_{\beta\alpha}/2)\alpha$. Thus, (R2) says that the projection of β onto α is a half-integer multiple of α.

Using the notion of a root system, one of the main results of the previous chapter can be reformulated as follows.

Theorem 7.3. *Let $\mathfrak{g}$ be a semisimple complex Lie algebra, with root decomposition (6.2). Then the set of roots $R \subset \mathfrak{h}_{\mathbb{R}}^* \setminus \{0\}$ is a reduced root system.*

Finally, for future use it is convenient to introduce, for every root $\alpha \in R$, the corresponding *coroot* $\alpha^\vee \in E^*$ defined by

$$\langle \alpha^\vee, \lambda \rangle = \frac{2(\alpha, \lambda)}{(\alpha, \alpha)}. \tag{7.4}$$

Note that for the root system of a semisimple Lie algebra, this coincides with the definition of $h_\alpha \in \mathfrak{h}$ defined by (6.5): $\alpha^\vee = h_\alpha$.

Then one easily sees that $\langle \alpha^\vee, \alpha \rangle = 2$ and that

$$\begin{aligned} n_{\alpha\beta} &= \langle \alpha, \beta^\vee \rangle \\ s_\alpha(\lambda) &= \lambda - \langle \lambda, \alpha^\vee \rangle \alpha. \end{aligned} \tag{7.5}$$

Example 7.4. Let e_i be the standard basis of $\mathbb{R}^n$, with the usual inner product: $(e_i, e_j) = \delta_{ij}$. Let $E = \{(\lambda_1, \ldots, \lambda_n) \in \mathbb{R}^n \mid \sum \lambda_i = 0\}$, and $R = \{e_i - e_j \mid 1 \le i, j \le n, i \ne j\} \subset E$. Then R is a reduced root system. Indeed, one easily sees that for $\alpha = e_i - e_j$, the corresponding reflection $s_\alpha : E \to E$ is the transposition of $i^{\text{th}}, j^{\text{th}}$ entries:

$$s_{e_i - e_j}(\ldots, \lambda_i, \ldots, \lambda_j, \ldots) = (\ldots, \lambda_j, \ldots, \lambda_i, \ldots)$$

Clearly, R is stable under such transpositions (and, more generally, under all permutations). Thus, condition (R3) is satisfied.

Since $(\alpha, \alpha) = 2$ for any $\alpha \in R$, condition (R2) is equivalent to $(\alpha, \beta) \in \mathbb{Z}$ for any $\alpha, \beta \in R$ which is immediate.

Finally, condition (R1) is obvious. Thus, R is a root system of rank $n - 1$. For historical reasons, this root system is usually referred to as "root system of type A_{n-1}" (subscript is chosen to match the rank of the root system).

Alternatively, one can also define E as a quotient of $\mathbb{R}^n$:

$$E = \mathbb{R}^n / \mathbb{R}(1, \ldots, 1).$$

In this description, we see that this root system is exactly the root system of Lie algebra $\mathfrak{sl}(n, \mathbb{C})$ (see Example 6.40).

7.2. Automorphisms and the Weyl group

Most important information about the root system is contained in the numbers $n_{\alpha\beta}$ rather than in inner product themselves. This motivates the following definition.

Definition 7.5. Let $R_1 \subset E_1, R_2 \subset E_2$ be two root systems. An isomorphism $\varphi \colon R_1 \to R_2$ is a vector space isomorphism $\varphi \colon E_1 \to E_2$ such that $\varphi(R_1) = R_2$ and $n_{\varphi(\alpha)\varphi(\beta)} = n_{\alpha\beta}$ for any $\alpha, \beta \in R_1$.

Note that condition $n_{\varphi(\alpha)\varphi(\beta)} = n_{\alpha\beta}$ will be automatically satisfied if φ preserves the inner product. However, not every isomorphism of root systems preserves the inner product. For example, for any $c \in \mathbb{R}_+$, the root systems R and $cR = \{c\alpha, \alpha \in R\}$ are isomorphic. The isomorphism is given by $v \mapsto cv$, which does not preserve the inner product.

A special class of automorphisms of a root system R are those generated by reflections s_α.

Definition 7.6. The Weyl group W of a root system R is the subgroup of $\mathrm{GL}(E)$ generated by reflections $s_\alpha, \alpha \in R$.

Lemma 7.7.

(1) *The Weyl group W is a finite subgroup in the orthogonal group $\mathrm{O}(E)$, and the root system R is invariant under the action of W.*
(2) *For any $w \in W, \alpha \in R$, we have $s_{w(\alpha)} = w s_\alpha w^{-1}$.*

Proof. Since every reflection s_α is an orthogonal transformation, $W \subset O(E)$. Since $s_\alpha(R) = R$ (by the axioms of a root system), we have $w(R) = R$ for any $w \in W$. Moreover, if some $w \in W$ leaves every root invariant, then $w = \text{id}$ (because R generates E). Thus, W is a subgroup of the group $\text{Aut}(R)$ of all automorphisms of R. Since R is a finite set, $\text{Aut}(R)$ is finite; thus W is also finite.

The second identity is obvious: indeed, $w s_\alpha w^{-1}$ acts as identity on the hyperplane $w L_\alpha = L_{w(\alpha)}$, and $w s_\alpha w^{-1}(w(\alpha)) = -w(\alpha)$, so it is a reflection corresponding to root $w(\alpha)$.   □

Example 7.8. Let R be the root system of type A_{n-1} (see Example 7.4). Then W is the group generated by transpositions s_{ij}. It is easy to see that these transpositions generate the symmetric group S_n; thus, for this root system $W = S_n$.

In particular, for the root system A_1 (i.e., the root system of $\mathfrak{sl}(2, \mathbb{C})$), we have $W = S_2 = \mathbb{Z}_2 = \{1, s\}$ where s acts on $E \simeq \mathbb{R}$ by $\lambda \mapsto -\lambda$.

It should be noted, however, that not all automorphisms of a root system are given by elements of the Weyl group. For example, for $A_n, n > 2$, the automorphism $\alpha \mapsto -\alpha$ is not in the Weyl group.

7.3. Pairs of roots and rank two root systems

Our main goal is to give a full classification of all possible reduced root systems, which in turn will be used to get a classification of all semisimple Lie algebras. The first step is considering the rank two case.

From now on, R is a reduced root system.

The first observation is that conditions (R2), (R3) impose very strong restrictions on relative position of two roots.

Theorem 7.9. *Let $\alpha, \beta \in R$ be roots which are not multiples of one another, with $|\alpha| \geq |\beta|$, and let φ be the angle between them. Then we must have one of the following possibilities:*

(1) $\varphi = \pi/2$ (*i.e.*, α, β *are orthogonal*), $n_{\alpha\beta} = n_{\beta\alpha} = 0$
(2a) $\varphi = 2\pi/3, |\alpha| = |\beta|, n_{\alpha\beta} = n_{\beta\alpha} = -1$
(2b) $\varphi = \pi/3, |\alpha| = |\beta|, n_{\alpha\beta} = n_{\beta\alpha} = 1.$
(3a) $\varphi = 3\pi/4, |\alpha| = \sqrt{2}|\beta|, n_{\alpha\beta} = -2, n_{\beta\alpha} = -1.$
(3b) $\varphi = \pi/4, \quad |\alpha| = \sqrt{2}|\beta|, n_{\alpha\beta} = 2, n_{\beta\alpha} = 1.$
(4a) $\varphi = 5\pi/6, |\alpha| = \sqrt{3}|\beta|, n_{\alpha\beta} = -3, n_{\beta\alpha} = -1.$
(4b) $\varphi = \pi/6, |\alpha| = \sqrt{3}|\beta|, n_{\alpha\beta} = 3, n_{\beta\alpha} = 1.$

Proof. Recall $n_{\alpha\beta}$ defined by (7.1). Since $(\alpha, \beta) = |\alpha||\beta|\cos\varphi$, we see that $n_{\alpha\beta} = 2\frac{|\alpha|}{|\beta|}\cos\varphi$. Thus, $n_{\alpha\beta}n_{\beta\alpha} = 4\cos^2\varphi$. Since $n_{\alpha\beta}n_{\beta\alpha} \in \mathbb{Z}$, this means that $n_{\alpha\beta}n_{\beta\alpha}$ must be one of $0, 1, 2, 3$. Analyzing each of these possibilities and using $\frac{n_{\alpha\beta}}{n_{\beta\alpha}} = \frac{|\alpha|^2}{|\beta|^2}$ if $\cos\varphi \neq 0$, we get the statement of the theorem. $\square$

It turns out that each of the possibilities listed in this theorem is indeed realized.

Theorem 7.10.

(1) *Let $A_1 \cup A_1, A_2, B_2, G_2$ be the sets of vectors in $\mathbb{R}^2$ shown in Figure 7.1. Then each of them is a rank two root system.*
(2) *Any rank two reduced root system is isomorphic to one of root systems $A_1 \cup A_1, A_2, B_2, G_2$.*

Proof. Proof of part (1) is given by explicit analysis. Since for any pair of vectors in these systems, the angle and ratio of lengths is among one of the possibilities listed in Theorem 7.9, condition (R2) is satisfied. It is also easy to see that condition (R3) is satisfied.

To prove the second part, assume that R is a reduced rank 2 root system. Let us choose α, β to be two roots such that the angle φ between them is as large as possible and $|\alpha| \geq |\beta|$. Then $\varphi \geq \pi/2$ (otherwise, we could take the pair $\alpha, s_\alpha(\beta)$ and get a larger angle). Thus, we must be in one of situations (1), (2a), (3a), (4a) of Theorem 7.9.

Consider, for example, case (2a): $|\alpha| = |\beta|, \varphi = 2\pi/3$. By the definition of a root system, R is stable under reflections s_α, s_β. But successively applying these two reflections to α, β we get exactly the root system of type A_2. Thus, in this case R contains as a subset the root system A_2 generated by α, β.

To show that in this case $R = A_2$, note that if we have another root γ which is not in A_2, then γ must be between some of the roots of A_2 (since R is reduced). Thus, the angle between γ and some root δ is less than $\pi/3$, and the angle between γ and $-\delta$ is greater than $2\pi/3$, which is impossible because the angle between α, β was chosen to be the maximal possible. Thus, $R = A_2$.

Similar analysis shows that in cases (1), (3a), (4a) of Theorem 7.9, we will get $R = A_1 \cup A_1, B_2, G_2$, respectively. $\square$

For future use, we also give the following result.

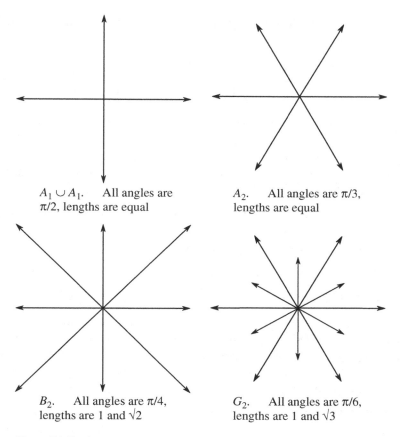

$A_1 \cup A_1$. All angles are $\pi/2$, lengths are equal

A_2. All angles are $\pi/3$, lengths are equal

B_2. All angles are $\pi/4$, lengths are 1 and $\sqrt{2}$

G_2. All angles are $\pi/6$, lengths are 1 and $\sqrt{3}$

Figure 7.1 Rank two root systems.

Lemma 7.11. *Let $\alpha, \beta \in R$ be two roots such that $(\alpha, \beta) < 0, \alpha \neq c\beta$. Then $\alpha + \beta \in R$.*

Proof. It suffices to prove this for each of rank two root systems described in Theorem 7.10. For each of them, it is easy to check directly. ☐

7.4. Positive roots and simple roots

In order to proceed with the classification of root systems, we would like to find for each root system some small set of "generating roots", similar to what was done in the previous section for rank 2 root systems. In general it can be done as follows.

Let $t \in E$ be such that for any root α, $(t, \alpha) \neq 0$ (such elements t are called *regular*). Then we can write

$$R = R_+ \sqcup R_-$$
$$R_+ = \{\alpha \in R \mid (\alpha, t) > 0\}, \quad R_- = \{\alpha \in R \mid (\alpha, t) < 0\}. \tag{7.6}$$

Such a decomposition will be called a *polarization* of R. Note that polarization depends on the choice of t. The roots $\alpha \in R_+$ will be called *positive*, and the roots $\alpha \in R_-$ will be called negative.

From now on, let us assume that we have fixed a polarization (7.6) of the root system R.

Definition 7.12. A root $\alpha \in R_+$ is called *simple* if it can not be written as a sum of two positive roots.

We will denote the set of simple roots by $\Pi \subset R_+$.

We have the following easy lemma.

Lemma 7.13. *Every positive root can be written as a sum of simple roots.*

Proof. If a positive root α is not simple, it can be written in the form $\alpha = \alpha' + \alpha''$, with $\alpha', \alpha'' \in R_+$, and $(\alpha', t) < (\alpha, t)$, $(\alpha'', t) < (\alpha, t)$. If α', α'' are not simple, we can apply the same argument to them to write them as a sum of positive roots. Since (α, t) can only take finitely many values, the process will terminate after finitely many steps. □

Example 7.14. Let us consider the root system A_2 and let t be as shown in Figure 7.2. Then there are three positive roots: two of them are denoted by α_1, α_2, and the third one is $\alpha_1 + \alpha_2$. Thus, one easily sees that α_1, α_2 are simple roots, and $\alpha_1 + \alpha_2$ is not simple.

Lemma 7.15. *If* $\alpha, \beta \in R_+$ *are simple,* $\alpha \neq \beta$ *then* $(\alpha, \beta) \leq 0$.

Proof. Assume that $(\alpha, \beta) > 0$. Then, applying Lemma 7.11 to $-\alpha, \beta$, we see that $\beta' = \beta - \alpha \in R$. If $\beta' \in R_+$, then $\beta = \beta' + \alpha$ can not be simple. If $\beta' \in R_-$, then $-\beta' \in R_+$, so $\alpha = -\beta' + \beta$ can not be simple. This contradiction shows that $(\alpha, \beta) > 0$ is impossible. □

Theorem 7.16. *Let* $R = R_+ \sqcup R_- \subset E$ *be a root system. Then the simple roots form a basis of the vector space E.*

Proof. By Lemma 7.13, every positive root can be written as a linear combination of simple roots. Since R spans E, this implies that the set of simple roots spans E.

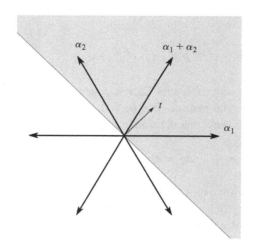

Figure 7.2 Positive and simple roots for A_2.

Linear independence of simple roots follows from the results of Lemma 7.15 and the following linear algebra lemma proof of which is given in the exercises (Exercise 7.3).

Lemma 7.17. *Let $v_1, \ldots v_k$ be a collection of non-zero vectors in a Euclidean space E such that for $i \neq j$, $(v_i, v_j) \leq 0$, $(v_i, t) > 0$ for some non-zero vector t. Then $\{v_i \ldots v_k\}$ are linearly independent.* $\square$

Corollary 7.18. *Every $\alpha \in R$ can be uniquely written as a linear combination of simple roots with integer coefficients:*

$$\alpha = \sum_{i=1}^{r} n_i \alpha_i, \qquad n_i \in \mathbb{Z}, \tag{7.7}$$

where $\{\alpha_1, \ldots, \alpha_r\} = \Pi$ is the set of simple roots. If $\alpha \in R_+$, then all $n_i \geq 0$; if $\alpha \in R_-$, then all $n_i \leq 0$.

For a positive root $\alpha \in R_+$, we define its *height* by

$$\mathrm{ht}\left(\sum n_i \alpha_i\right) = \sum n_i \in \mathbb{Z}_+, \tag{7.8}$$

so that $\mathrm{ht}(\alpha_i) = 1$. In many cases, statements about positive roots can be proved by induction in height.

Example 7.19. Let R be the root system of type A_{n-1} or equivalently, the root system of $\mathfrak{sl}(n, \mathbb{C})$ (see Example 6.40, Example 7.4). Choose the polarization as follows:

$$R_+ = \{e_i - e_j \mid i < j\}$$

(the corresponding root subspaces $E_{ij}, i < j$, generate the Lie subalgebra $\mathfrak{n}$ of strictly upper-triangular matrices in $\mathfrak{sl}(n, \mathbb{C})$).

Then it is easy to show that the simple roots are

$$\alpha_1 = e_1 - e_2, \qquad \alpha_2 = e_2 - e_3, \qquad \ldots, \qquad \alpha_{n-1} = e_{n-1} - e_n,$$

and indeed, any positive root can be written as a sum of simple roots with non-negative integer coefficients. For example, $e_2 - e_4 = (e_2 - e_3) + (e_3 - e_4) = \alpha_2 + \alpha_3$. The height is given by $\mathrm{ht}(e_i - e_j) = j - i$.

7.5. Weight and root lattices

In the study of root systems of simple Lie algebras, we will frequently use the following lattices. Recall that a lattice in a real vector space E is an abelian group generated by a basis in E. Of course, by a suitable change of basis any lattice $L \subset E$ can be identified with $\mathbb{Z}^n \subset \mathbb{R}^n$.

Every root system $R \subset E$ gives rise to the following lattices:

$$\begin{aligned} Q &= \{\text{abelian group generated by } \alpha \in R\} \subset E \\ Q^\vee &= \{\text{abelian group generated by } \alpha^\vee, \alpha \in R\} \subset E^* \end{aligned} \tag{7.9}$$

Lattice Q is called the *root lattice* of R, and $Q^\vee$ is the *coroot lattice*. Note that despite the notation, $Q^\vee$ is **not** the dual lattice to Q.

To justify the use of the word lattice, we need to show that $Q, Q^\vee$ are indeed generated by a basis in E (respectively E^*). This can be done as follows. Fix a polarization of R and let $\Pi = \{\alpha_1, \ldots, \alpha_r\}$ be the corresponding system of simple roots. Since every root can be written as a linear combination of simple roots with integer coefficients (Corollary 7.18), one has

$$Q = \bigoplus \mathbb{Z}\alpha_i, \tag{7.10}$$

which shows that Q is indeed a lattice. Similarly, it follows from Exercise 7.2 that

$$Q^\vee = \bigoplus \mathbb{Z}\alpha_i^\vee. \tag{7.11}$$

Even more important in the applications to the representation theory of semisimple Lie algebras is the *weight lattice* $P \subset E$ defined as follows:

$$P = \{\lambda \in E \mid \langle \lambda, \alpha^\vee \rangle \in \mathbb{Z} \text{ for all } \alpha \in R\}$$
$$= \{\lambda \in E \mid \langle \lambda, \alpha^\vee \rangle \in \mathbb{Z} \text{ for all } \alpha^\vee \in Q^\vee\}. \tag{7.12}$$

In other words, $P \subset E$ is exactly the dual lattice of $Q^\vee \subset E^*$. Elements of P are frequently called *integral* weights. Their role in representation theory will be discussed in Chapter 8.

Since $Q^\vee$ is generated by $\alpha_i^\vee$, the weight lattice can also be defined by

$$P = \{\lambda \in E \mid \langle \lambda, \alpha_i^\vee \rangle \in \mathbb{Z} \text{ for all simple roots } \alpha_i\}. \tag{7.13}$$

One can easily define a basis in P. Namely, define *fundamental weights* $\omega_i \in E$ by

$$\langle \omega_i, \alpha_j^\vee \rangle = \delta_{ij}. \tag{7.14}$$

Then one easily sees that so defined ω_i form a basis in E and that

$$P = \bigoplus_i \mathbb{Z}\omega_i.$$

Finally, note that by the axioms of a root system, we have $n_{\alpha\beta} = \langle \alpha, \beta^\vee \rangle \in \mathbb{Z}$ for any roots α, β. Thus, $R \subset P$ which implies that

$$Q \subset P.$$

However, in general $P \neq Q$, as the examples below show. Since both P, Q are free abelian groups of rank r, general theory of finitely generated abelian groups implies that the quotient group P/Q is a finite abelian group. It is also possible to describe the order $|P/Q|$ in terms of the matrix $a_{ij} = \langle \alpha_i^\vee, \alpha_j \rangle$ (see Exercise 7.4).

Example 7.20. Consider the root system A_1. It has the unique positive root α, so $Q = \mathbb{Z}\alpha, Q^\vee = \mathbb{Z}\alpha^\vee$. If we define the inner product $(\ ,\ )$ by $(\alpha, \alpha) = 2$, and use this product to identify $E^* \simeq E$, then under this identification $\alpha^\vee \mapsto \alpha, Q^\vee \xrightarrow{\sim} Q$.

Since $\langle \alpha, \alpha^\vee \rangle = 2$, we see that the fundamental weight is $\omega = \alpha/2$, and $P = \mathbb{Z}(\alpha/2)$. Thus, in this case $P/Q = \mathbb{Z}_2$.

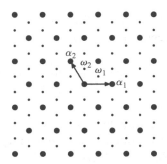

Figure 7.3 Weight and root lattices for A_2. Large dots show $\alpha \in Q$, small dots $\alpha \in P - Q$.

Example 7.21. For the root system A_2, the root and weight lattices are shown in Figure 7.3. This figure also shows simple roots α_1, α_2 and fundamental weights ω_1, ω_2.

It is easy to see from the figure (and also easy to prove algebraically) that one can take α_1, ω_1 as a basis of P, and that $\alpha_1, 3\omega_1 = \alpha_2 + 2\alpha_1$ is a basis of Q. Thus, $P/Q = \mathbb{Z}_3$.

7.6. Weyl chambers

In the previous sections, we have constructed, starting with a root system R, first the set of positive roots R_+ and then a smaller set of simple roots $\Pi = \{\alpha_1, \ldots, \alpha_r\}$ which in a suitable sense generates R. Schematically this can be shown as follows:

$$R \longrightarrow R_+ \longrightarrow \Pi = \{\alpha_1, \ldots, \alpha_r\}.$$

The first step (passage from R to R_+) requires a choice of polarization, which is determined by a regular element $t \in E$; the second step is independent of any choices.

Our next goal is to use this information to get a classification of reduced root systems, by classifying possible sets of simple roots. However, before doing this we need to answer the following two questions:

(1) Is it possible to recover R from Π?
(2) Do different choices of polarization give rise to equivalent in a suitable sense sets of simple roots Π, Π'?

We will start with the second question as it is easier to answer.

Recall that a polarization is defined by an element $t \in E$, which does not lie on any of the hyperplanes orthogonal to roots:

$$t \in E \setminus \bigcup_{\alpha \in R} L_\alpha$$

$$L_\alpha = \{\lambda \in E \mid (\alpha, \lambda) = 0\}.$$

(7.15)

Moreover, the polarization actually depends not on t itself but only on the signs of (t, α); thus, polarization is unchanged if we change t as long as we do not cross any of the hyperplanes. This justifies the following definition.

Definition 7.22. A Weyl chamber is a connected component of the complement to the hyperplanes:

$$C = \text{connected component of } \left(E \setminus \bigcup_{\alpha \in R} L_\alpha \right).$$

For example, for root system A_2 there are six Weyl chambers; one of them is shaded in Figure 7.4.

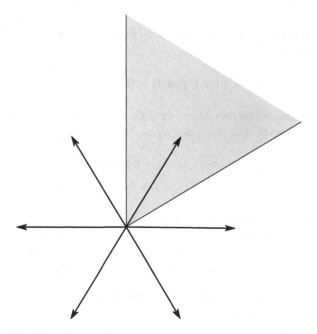

Figure 7.4 A Weyl chamber for A_2.

Clearly, to specify a Weyl chamber we need to specify, for each hyperplane L_α, on which side of the hyperplane the Weyl chamber lies. Thus, a Weyl chamber is defined by a system of inequalities of the form

$$\pm(\alpha, \lambda) > 0$$

(one inequality for each root hyperplane). Any such system of inequalities defines either an empty set or a Weyl chamber.

For future use, we state here some results about the geometry of the Weyl chambers.

Lemma 7.23.

(1) *The closure $\overline{C}$ of a Weyl chamber C is an unbounded convex cone.*
(2) *The boundary $\partial \overline{C}$ is a union of finite number of codimension one faces:*
    $\partial \overline{C} = \bigcup F_i$. *Each F_i is a closed convex unbounded subset in one of the hyperplanes L_α, given by a system of inequalities. The hyperplanes containing F_i are called* walls *of C.*

This lemma is geometrically obvious (in fact, it equally applies to any subset in a Euclidean space defined by a finite system of strict inequalities) and we omit the proof.

We can now return to the polarizations. Note that any Weyl chamber C defines a polarization given by

$$R_+ = \{\alpha \in R \mid (\alpha, t) > 0\}, \qquad t \in C \qquad (7.16)$$

(this does not depend on the choice of $t \in C$). Conversely, given a polarization $R = R_+ \sqcup R_-$, define the corresponding *positive Weyl chamber C_+* by

$$\begin{aligned} C_+ &= \{\lambda \in E \mid (\lambda, \alpha) > 0 \text{ for all } \alpha \in R_+\} \\ &= \{\lambda \in E \mid (\lambda, \alpha_i) > 0 \text{ for all } \alpha_i \in \Pi\} \end{aligned} \qquad (7.17)$$

(to prove the last equality, note that if $(\lambda, \alpha_i) > 0$ for all $\alpha_i \in \Pi$, then by Lemma 7.13, for any $\alpha = \sum n_i \alpha_i$, we have $(\lambda, \alpha) > 0$). This system of inequalities does have solutions (because the element t used to define the polarization satisfies these inequalities) and thus defines a Weyl chamber.

Lemma 7.24. *Formulas (7.16), (7.17) define a bijection between the set of all polarizations of R and the set of Weyl chambers.*

Proof is left as an exercise to the reader.

In order to relate polarizations defined by different Weyl chambers, recall the Weyl group W defined in Section 7.2 Since the action of W maps root hyperplanes to root hyperplanes, we have a well-defined action of W on the set of Weyl chambers.

Theorem 7.25. *The Weyl group acts transitively on the set of Weyl chambers.*

Proof. The proof is based on several facts which are of significant interest in their own right. Namely, let us say that two Weyl chambers C, C' are *adjacent* if they have a common codimension one face F (obviously, they have to be on different sides of F). If L_α is the hyperplane containing this common face F, then we will say that C, C' are adjacent chambers separated by L_α.

Then we have the following two lemmas, proof of which as an exercise to the reader.

Lemma 7.26. *Any two Weyl chambers C, C' can be connected by a sequence of chambers $C_0 = C, C_1, \ldots, C_l = C'$ such that C_i is adjacent to C_{i+1}.*

Lemma 7.27. *If C, C' are adjacent Weyl chambers separated by hyperplane L_α then $s_\alpha(C) = C'$.*

The statement of the theorem now easily follows from these two lemmas. Indeed, let C, C' be two Weyl chambers. By Lemma 7.26, they can be connected by a sequence of Weyl chambers $C_0 = C, C_1, \ldots, C_l = C'$. Let L_{β_i} be the hyperplane separating C_{i-1} and C_i. Then, by Lemma 7.27,

$$C_l = s_{\beta_l}(C_{l-1}) = s_{\beta_l} s_{\beta_{l-1}}(C_{l-2}) = \ldots$$
$$= s_{\beta_l} \ldots s_{\beta_1}(C_0)$$

(7.18)

so $C' = w(C)$, with $w = s_{\beta_l} \ldots s_{\beta_1}$. This completes the proof of Theorem 7.25.

□

Corollary 7.28. *Every Weyl chamber has exactly $r = \operatorname{rank}(R)$ walls. Walls of positive Weyl chamber C_+ are $L_{\alpha_i}, \alpha_i \in \Pi$.*

Proof. For the positive Weyl chamber C_+, this follows from (7.17). Since every Weyl chamber can be written in the form $C = w(C_+)$ for some $w \in W$, all Weyl chambers have the same number of walls.

□

Corollary 7.29. *Let $R = R_+ \sqcup R_- = R'_+ \sqcup R'_-$ be two polarizations of the same root system, and Π, Π' the corresponding sets of simple roots. Then there exists an element $w \in W$ such that $\Pi = w(\Pi')$.*

Proof. By Lemma 7.24, each polarization is defined by a Weyl chamber. Since W acts transitively on the set of Weyl chambers, it also acts transitively on the set of all polarizations. $\square$

This last corollary provides an answer to the question asked in the beginning of this section: sets of simple roots obtained from different polarizations can be related by an orthogonal transformation of E.

7.7. Simple reflections

We can now return to the first question asked in the beginning of the previous section: is it possible to recover R from the set of simple roots Π? The answer is again based on the use of Weyl group.

Theorem 7.30. *Let R be a reduced root system, with fixed polarization $R = R_+ \sqcup R_-$. Let $\Pi = \{\alpha_1, \ldots, \alpha_r\}$ be the set of simple roots. Consider reflections corresponding to simple roots $s_i = s_{\alpha_i}$ (they are called* simple reflections*).*

(1) *The simple reflections s_i generate W.*
(2) *$W(\Pi) = R$: every $\alpha \in R$ can be written in the form $w(\alpha_i)$ for some $w \in W$ and $\alpha_i \in \Pi$.*

Proof. We start by proving the following result

Lemma 7.31. *Any Weyl chamber can be written as*

$$C = s_{i_1} \ldots s_{i_l}(C_+)$$

for some sequence of indices $i_1, \ldots, i_l \in \{1, \ldots, r\}$. (Here C_+ is the positive Weyl chamber defined by (7.17).)

Proof. By the construction given in the proof of Theorem 7.25, we can connect C_+, C by a chain of adjacent Weyl chambers $C_0 = C_+, C_1, \ldots, C_l = C$. Then $C = s_{\beta_l} \ldots s_{\beta_1}(C_+)$, where L_{β_i} is the hyperplane separating C_{i-1} and C_i.

Since L_{β_1} separates $C_0 = C_+$ from C_1, it means that L_{β_1} is one of the walls of C_+. Since the walls of C_+ are exactly hyperplanes L_{α_i} corresponding to simple roots (see Corollary 7.28), we see that $\beta_1 = \pm\alpha_{i_1}$ for some index $i_1 \in \{1, \ldots, r\}$, so $s_{\beta_1} = s_{i_1}$ and $C_1 = s_{i_1}(C_+)$.

Consider now the hyperplane L_{β_2} separating C_1 from C_2. It is a wall of $C_1 = s_{i_1}(C_+)$; thus, it must be of the form $L_{\beta_2} = s_{i_1}(L)$ for some hyperplane L which is a wall of C_+. Thus, we get that $\beta_2 = \pm s_{i_1}(\alpha_{i_2})$ for some index i_2.

By Lemma 7.7, we therefore have $s_{\beta_2} = s_{i_1} s_{i_2} s_{i_1}$ and thus

$$s_{\beta_2} s_{\beta_1} = s_{i_1} s_{i_2} s_{i_1} \cdot s_{i_1} = s_{i_1} s_{i_2}$$
$$C_2 = s_{i_1} s_{i_2}(C_+).$$

Repeating the same argument, we finally get that

$$C = s_{i_1} \ldots s_{i_l}(C_+)$$

and the indices i_k are computed inductively, by

$$\beta_k = s_{i_1} \ldots s_{i_{k-1}}(\alpha_{i_k}) \tag{7.19}$$

which completes the proof of the lemma. $\qquad\square$

Now the theorem easily follows. Indeed, every hyperplane L_α is a wall of some Weyl chamber C. Using the lemma, we can write $C = w(C_+)$ for some $w = s_{i_1} \ldots s_{i_l}$. Thus, $L_\alpha = w(L_{\alpha_j})$ for some index j, so $\alpha = \pm w(\alpha_j)$ and $s_\alpha = w s_j w^{-1}$, which proves both statements of the theorem. $\qquad\square$

It is also possible to write the full set of defining relations for W (see Exercise 7.11).

Example 7.32. Let R be the root system of type A_{n-1}. Then the Weyl group is $W = S_n$ (see Example 7.8) and simple reflections are transpositions $s_i = (i\ i+1)$. And indeed, it is well known that these transpositions generate the symmetric group.

We can also describe the Weyl chambers in this case. Namely, the positive Weyl chamber is

$$C_+ = \{(\lambda_1, \ldots, \lambda_n) \in E \mid \lambda_1 < \lambda_2 < \ldots < \lambda_n\}$$

and all other Weyl chambers are obtained by applying to C_+ permutations $\sigma \in S_n$. Thus, they are of the form

$$C_\sigma = \{(\lambda_1, \ldots, \lambda_n) \in E \mid \lambda_{\sigma(1)} < \lambda_{\sigma(2)} < \ldots < \lambda_{\sigma(n)}\}, \quad \sigma \in S_n$$

Corollary 7.33. *The root system R can be recovered from the set of simple roots Π.*

Proof. Given Π, we can recover W as the group generated by s_i and then recover $R = W(\Pi)$. $\qquad\square$

Let us say that a root hyperplane L_α *separates* two Weyl chambers C, C' if these two chambers are on different sides of L_α, i.e. $\alpha(C), \alpha(C')$ have different signs (we do not assume that L_α is one of the walls of C or C').

Definition 7.34. Let R be a reduced root system, with the set of simple roots Π. Then we define, for an element $w \in W$, its *length* by

$$l(w) = \text{number of root hyperplanes separating } C_+ \text{ and } w(C_+)$$
$$= |\{\alpha \in R_+ \mid w(\alpha) \in R_-\}|. \tag{7.20}$$

It should be noted that $l(w)$ depends not only on w itself but also on the choice of polarization $R = R_+ \sqcup R_-$ or equivalently, the set of simple roots.

Example 7.35. Let $w = s_i$ be a simple reflection. Then the Weyl chambers C_+ and $s_i(C_+)$ are separated by exactly one hyperplane, namely L_{α_i}. Therefore, $l(s_i) = 1$, and

$$\{\alpha \in R_+ \mid s_i(\alpha) \in R_-\} = \{\alpha_i\}. \tag{7.21}$$

In other words, $s_i(\alpha_i) = -\alpha_i \in R_-$ and s_i permutes elements of $R_+ \setminus \{\alpha_i\}$.

This example is very useful in many arguments involving Weyl group, such as the following lemma.

Lemma 7.36. *Let*

$$\rho = \frac{1}{2} \sum_{\alpha \in R_+} \alpha. \tag{7.22}$$

Then $\langle \rho, \alpha_i^\vee \rangle = 2(\rho, \alpha_i)/(\alpha_i, \alpha_i) = 1$.

Proof. Writing $\rho = (\alpha_i + \sum_{\alpha \in R_+ \setminus \{\alpha_i\}} \alpha)/2$ and using results of Example 7.35, we see that $s_i(\rho) = \rho - \alpha_i$. On the other hand, by definition $s_i(\lambda) = \lambda - \langle \alpha_i^\vee, \lambda \rangle \alpha_i$. $\qquad\square$

Theorem 7.37. *Let*

$$w = s_{i_1} \dots s_{i_l}$$

be an expression for w as a product of simple reflections which has minimal possible length (such expressions are called reduced*). Then $l = l(w)$.*

Proof. We can connect C_+ and $w(C_+)$ by a chain of Weyl chambers $C_0 = C_+, C_1, \dots, C_l = w(C_+)$, where $C_k = s_{i_1} \dots s_{i_k}(C_+)$. The same argument as in the proof of Theorem 7.31 shows that then C_k and C_{k-1} are adjacent Weyl chambers separated by root hyperplane L_{β_k}, with $\beta_k = s_{i_1} \dots s_{i_{k-1}}(\alpha_{i_k})$. This shows that we can connect C_+ and $w(C_+)$ by a path crossing exactly l

hyperplanes. In particular, this means that C_+ and $w(C_+)$ are separated by at most l hyperplanes, so $l(w) \leq l$.

Note, however, that we can not yet conclude that $l(w) = l$: it is possible that the path we had constructed crosses some hyperplane more than once. For example, we can write $1 = s_i s_i$, which gives us a path connecting C_+ with itself but crossing hyperplane L_{α_i} twice. So to show that $l(w) = l$, we need to show that if $w = s_{i_1} \ldots s_{i_l}$ is a **reduced** expression, then all hyperplanes $L_{\beta_1}, \ldots, L_{\beta_l}$ are distinct: we never cross any hyperplane more than once. The proof of this fact is given as an exercise (see Exercise 7.6). $\square$

Corollary 7.38. *The action of W on the set of Weyl chambers is simply transitive.*

Proof. Otherwise, there exists $w \in W$ such that $w(C_+) = C_+$. By definition, this means $l(w) = 0$. By Theorem 7.37, this implies that $w = 1$. $\square$

This shows that $\overline{C}_+$ is the fundamental domain for the action of W on E. In fact, we have a stronger result: every W-orbit in E contains exactly one element from $\overline{C}_+$ (see Exercise 7.8).

Lemma 7.39. *Let C_- be the negative Weyl chamber: $C_- = -C_+$ and let $w_0 \in W$ be such that $w_0(C_+) = C_-$ (by Corollary 7.38, such an element exists and is unique). Then $l(w_0) = |R_+|$ and for any $w \in W, w \neq w_0$, we have $l(w) < l(w_0)$. For this reason w_0 is called the* longest element *in W.*

The proof of this lemma is left to the reader as an exercise.

7.8. Dynkin diagrams and classification of root systems

In the previous sections, we have discussed that given a reduced root system R, we can choose a polarization $R = R_+ \sqcup R_-$ and then define the set of simple roots $\Pi = \{\alpha_1, \ldots, \alpha_r\}$. We have shown that R can be recovered from Π (Corollary 7.33) and that different choices of polarization give rise to sets of simple roots which are related by the action of the Weyl group (Corollary 7.29). Thus, classifying root systems is equivalent to classifying possible sets of simple roots Π.

The main goal of this section will be to give a complete solution of this problem, thus giving a classification of all root systems.

The first step is to note that there is an obvious way to construct larger root systems from smaller ones. Namely, if $R_1 \subset E_1$ and $R_2 \subset E_2$ are two root systems, then we can define $R = R_1 \sqcup R_2 \subset E_1 \oplus E_2$, with the inner product on $E_1 \oplus E_2$ defined so that $E_1 \perp E_2$. It is easy to see that so defined R is again a root system.

Definition 7.40. A root system R is called *reducible* if it can be written in the form $R = R_1 \sqcup R_2$, with $R_1 \perp R_2$. Otherwise, R is called *irreducible*.

For example, the root system $A_1 \cup A_1$ discussed in Section 7.3 is reducible; all other root systems discussed in that section are irreducible.

Remark 7.41. It should be noted that a root system being reducible or irreducible is completely unrelated to whether the root system is reduced or not. It would be best if a different terminology were used, to avoid confusion; however, both of these terms are so widely used that changing them is not feasible.

There is an analogous notion for the set of simple roots.

Lemma 7.42. *Let R be a reduced root system, with given polarization, and let Π be the set of simple roots.*

(1) *If R is reducible: $R = R_1 \sqcup R_2$, then $\Pi = \Pi_1 \sqcup \Pi_2$, where $\Pi_i = \Pi \cap R_i$ is the set of simple roots for R_i.*
(2) *Conversely, if $\Pi = \Pi_1 \sqcup \Pi_2$, with $\Pi_1 \perp \Pi_2$, then $R = R_1 \sqcup R_2$, where R_i is the root system generated by Π_i.*

Proof. The first part is obvious. To prove the second part, notice that if $\alpha \in \Pi_1, \beta \in \Pi_2$, then $s_\alpha(\beta) = \beta$ and s_α, s_β commute. Thus, if we denote by W_i the group generated by simple reflections $s_\alpha, \alpha \in \Pi_i$, then $W = W_1 \times W_2$, and W_1 acts trivially on Π_2, W_2 acts trivially on Π_1. Thus, $R = W(\Pi_1 \sqcup \Pi_2) = W_1(\Pi_1) \sqcup W_2(\Pi_2)$. $\qquad\qquad\square$

It can be shown that every reducible root system can be uniquely written in the form $R_1 \sqcup R_2 \cdots \sqcup R_n$, where R_i are mutually orthogonal irreducible root systems. Thus, in order to classify all root systems, it suffices to classify all irreducible root systems. For this reason, from now on R is an irreducible root system and Π is the corresponding set of simple roots. We assume that we have chosen an order on the set of simple roots: $\Pi = \{\alpha_1 \ldots, \alpha_r\}$.

The compact way of describing relative position of roots $\alpha_i \in \Pi$ is by writing all inner products between these roots. However, this is not invariant under isomorphisms of root systems. A better way of describing the relative position of simple roots is given by the so called Cartan matrix.

Definition 7.43. The Cartan matrix A of a set of simple roots $\Pi \subset R$ is the $r \times r$ matrix with entries

$$a_{ij} = n_{\alpha_j \alpha_i} = \langle \alpha_i^\vee, \alpha_j \rangle = \frac{2(\alpha_i, \alpha_j)}{(\alpha_i, \alpha_i)}. \tag{7.23}$$

The following properties of the Cartan matrix immediately follow from the definitions and from known properties of simple roots.

Lemma 7.44.

(1) For any $i, a_{ii} = 2$.
(2) For any $i \neq j, a_{ij}$ is a non-positive integer: $a_{ij} \in \mathbb{Z}, a_{ij} \leq 0$.
(3) For any $i \neq j, a_{ij} a_{ji} = 4 \cos^2 \varphi$, where φ is the angle between α_i, α_j. If $\varphi \neq \pi/2$, then

$$\frac{|\alpha_i|^2}{|\alpha_j|^2} = \frac{a_{ji}}{a_{ij}}.$$

Example 7.45. For the root system A_n, the Cartan matrix is

$$A = \begin{bmatrix} 2 & -1 & & & & \\ -1 & 2 & -1 & & & \\ & -1 & 2 & -1 & & \\ & & & \vdots & & \\ & & & -1 & 2 & -1 \\ & & & & -1 & 2 \end{bmatrix}$$

(entries which are not shown are zeroes).

The information contained in the Cartan matrix can also be presented in a graphical way.

Definition 7.46. Let Π be a set of simple roots of a root system R. The *Dynkin diagram* of Π is the graph constructed in the following manner.

- For each simple root α_i, we construct a vertex v_i of the Dynkin diagram (traditionally, vertices are drawn as small circles rather than as dots).
- For each pair of simple roots $\alpha_i \neq \alpha_j$, we connect the corresponding vertices by n edges, where n depends on the angle φ between α_i, α_j:
  For $\varphi = \pi/2, n = 0$ (vertices are not connected)
  For $\varphi = 2\pi/3, n = 1$ (case of A_2 system)
  For $\varphi = 3\pi/4, n = 2$ (case of B_2 system)
  For $\varphi = 5\pi/6, n = 3$ (case of G_2 system).

- Finally, for every pair of distinct simple roots $\alpha_i \neq \alpha_j$, if $|\alpha_i| \neq |\alpha_j|$ and they are not orthogonal, we orient the corresponding (multiple) edge by putting on it an arrow pointing towards the shorter root.

Example 7.47. The Dynkin diagrams for rank two root systems are shown in Figure 7.5.

Theorem 7.48. *Let Π be a set of simple roots of a reduced root system R.*

(1) *The Dynkin diagram of Π is connected if and only if R is irreducible.*
(2) *The Dynkin diagram determines the Cartan matrix A.*
(3) *R is determined by the Dynkin diagram uniquely up to an isomorphism: if R, R' are two reduced root systems with the same Dynkin diagram, then they are isomorphic.*

Proof. (1) Assume that R is reducible; then, by Lemma 7.42, we have $\Pi = \Pi_1 \sqcup \Pi_2$, with $\Pi_1 \perp \Pi_2$. Thus, by construction of Dynkin diagram, it will be a disjoint union of the Dynkin diagram of Π_1 and the Dynkin diagram of Π_2. Proof in the opposite direction is similar.

(2) Dynkin diagram determines, for each pair of simple roots α_i, α_j, the angle between them and shows which of them is longer. Since all possible configurations of two roots are listed in Theorem 7.9, one easily sees that this information, together with the condition $(\alpha_i, \alpha_j) \leq 0$, uniquely determines $n_{\alpha_i \alpha_j}, n_{\alpha_j \alpha_i}$.

(3) By part (2), the Dynkin diagram determines Π uniquely up to an isomorphism. By Corollary 7.33, Π determines R uniquely up to an isomorphism. $\qquad \square$

Thus, the problem of classifying all irreducible root systems reduces to the following problem: which graphs can appear as Dynkin diagrams of irreducible root systems? The answer is given by the following theorem.

Theorem 7.49. *Let R be a reduced irreducible root system. Then its Dynkin diagram is isomorphic to one of the diagrams below (in each diagram, the subscript is equal to the number of vertices, so X_n has exactly n vertices):*

Figure 7.5 Dynkin diagrams of rank two root systems.

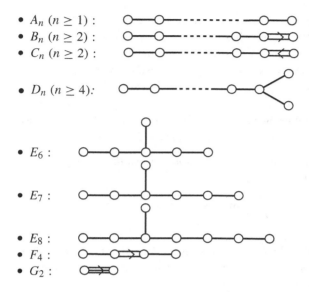

- A_n $(n \geq 1)$:
- B_n $(n \geq 2)$:
- C_n $(n \geq 2)$:

- D_n $(n \geq 4)$:

- E_6 :

- E_7 :

- E_8 :
- F_4 :
- G_2 :

Conversely, each of these diagrams does appear as the Dynkin diagram of a reduced irreducible root system.

The proof of this theorem is not difficult but rather long as it requires analyzing a number of cases. We will give a proof of a special case, when the diagram contains no multiple edges, in Section 7.10

Explicit constructions of the root systems corresponding to each of the diagrams A–D is given in Appendix A, along with useful information such as a description of the Weyl group, and much more. A description of root systems E_6, E_7, E_8, F_4, G_2 (these root systems are sometimes called "exceptional") can be found in [3, 24].

The letters $A, B, \ldots, G$ do not have any deep significance: these are just the first seven letters of the alphabet. However, this notation has become standard. Since the Dynkin diagram determines the root system up to isomorphism, it is also common to use the same notation $A_n, \ldots, G_2$ for the corresponding root system.

Remark 7.50. In the list above, we have imposed restrictions $n \geq 2$ for B_n, C_n and $n \geq 4$ for D_n, which is natural from the point of view of diagrams. In fact, constructions of root systems given in Appendix A also work for smaller values of n; namely, constructions of root systems B_n, C_n also make sense for $n = 1$; however, these root systems coincide with A_1 : $B_1 = C_1 = A_1$, so they do not give any new diagrams (note also that for $n = 2, B_2 = C_2$). (This corresponds to Lie algebra isomorphisms $\mathfrak{sl}(2, \mathbb{C}) \simeq \mathfrak{so}(3, \mathbb{C}) \simeq \mathfrak{sp}(1, \mathbb{C})$ constructed in Section 3.10 and $\mathfrak{sp}(2, \mathbb{C}) \simeq \mathfrak{so}(5, \mathbb{C})$.)

Similarly, construction of the root system D_n also makes sense for $n = 2, 3$, in which case it gives $D_2 = A_1 \cup A_1, D_3 = A_3$, which correspond to Lie algebra isomorphisms $\mathfrak{so}(4, \mathbb{C}) \simeq \mathfrak{sl}(2, \mathbb{C}) \oplus \mathfrak{sl}(2, \mathbb{C}), \mathfrak{so}(6, \mathbb{C}) \simeq \mathfrak{sl}(4, \mathbb{C})$, see Exercise 6.6.

Other than the equalities listed above, all root systems $A_n, \ldots, G_2$ are distinct.

Corollary 7.51. *If R is a reduced irreducible root system, then (α, α) can take at most two different values. The number*

$$m = \frac{\max(\alpha, \alpha)}{\min(\alpha, \alpha)} \tag{7.24}$$

is equal to the maximal multiplicity of an edge in the Dynkin diagram; thus, $m = 1$ for root systems of types ADE (these are called simply-laced *diagrams), $m = 2$ for types BCF, and $m = 3$ for G_2.*

For non-simply laced systems, the roots with (α, α) being the larger of two possible values are called the long *roots, and the remaining roots are called* short.

7.9. Serre relations and classification of semisimple Lie algebras

We can now return to the question of classification of complex semisimple Lie algebras. Since every semisimple algebra is a direct sum of simple ones, it suffices to classify simple Lie algebras.

According to the results of Section 6.6, every semisimple Lie algebra defines a reduced root system; if the algebra is not simple but only semisimple, then the root system is reducible. The one question we have not yet answered is whether one can go back and recover the Lie algebra from the root system. If the answer is positive, then the isomorphism classes of simple Lie algebras are in bijection with the isomorphism classes of reduced irreducible root systems, and thus we could use classification results of Section 7.8 to classify simple Lie algebras.

Theorem 7.52. *Let $\mathfrak{g}$ be a semisimple complex Lie algebra with root system $R \subset \mathfrak{h}^*$, and $(\ ,\ ) - a$ non-degenerate invariant symmetric bilinear form on $\mathfrak{g}$. Let $R = R_+ \sqcup R_-$ be a polarization of R, and $\Pi = \{\alpha_1, \ldots, \alpha_r\} -$ the corresponding system of simple roots.*

(1) *The subspaces*

$$\mathfrak{n}_{\pm} = \bigoplus_{\alpha \in R_{\pm}} \mathfrak{g}_{\alpha} \tag{7.25}$$

are subalgebras in $\mathfrak{g}$, *and*

$$\mathfrak{g} = \mathfrak{n}_- \oplus \mathfrak{h} \oplus \mathfrak{n}_+ \tag{7.26}$$

as a vector space.

(2) *Let* $e_i \in \mathfrak{g}_{\alpha_i}, f_i \in \mathfrak{g}_{-\alpha_i}$ *be chosen so that* $(e_i, f_i) = 2/(\alpha_i, \alpha_i)$, *and let* $h_i = h_{\alpha_i} \in \mathfrak{h}$ *be defined by* (6.5). *Then* $e_1, \ldots, e_r$ *generate* $\mathfrak{n}_+, f_1, \ldots, f_r$ *generate* $\mathfrak{n}_-$, *and* $h_1, \ldots, h_r$ *form a basis of* $\mathfrak{h}$. *In particular,* $\{e_i, f_i, h_i\}_{i=1\ldots r}$ *generate* $\mathfrak{g}$.

(3) *The elements* e_i, f_i, h_i *satisfy the following relations, called the* Serre *relations:*

$$[h_i, h_j] = 0 \tag{7.27}$$

$$[h_i, e_j] = a_{ij} e_j, \qquad [h_i, f_j] = -a_{ij} f_j \tag{7.28}$$

$$[e_i, f_j] = \delta_{ij} h_i \tag{7.29}$$

$$(\text{ad } e_i)^{1-a_{ij}} e_j = 0 \qquad i \neq j \tag{7.30}$$

$$(\text{ad } f_i)^{1-a_{ij}} f_j = 0 \qquad i \neq j \tag{7.31}$$

where $a_{ij} = n_{\alpha_j, \alpha_i} = \langle \alpha_i^\vee, \alpha_j \rangle$ *are the entries of the Cartan matrix.*

Proof.

(1) The fact that $\mathfrak{n}_+$ is a subalgebra follows from $[\mathfrak{g}_\alpha, \mathfrak{g}_\beta] \subset \mathfrak{g}_{\alpha+\beta}$ (see Theorem 6.38) and the fact that the sum of positive roots is positive. Equation (7.26) is obvious.

(2) The fact that h_i form a basis of $\mathfrak{h}$ follows from Theorem 7.16. To prove that e_i generate $\mathfrak{n}_+$, we first prove the following lemma.

Lemma 7.53. *Let* $R = R_+ \sqcup R_-$ *be a reduced root system, with a set of simple roots* $\{\alpha_1, \ldots, \alpha_r\}$. *Let* α *be a positive root which is not simple. Then* $\alpha = \beta + \alpha_i$ *for some positive root* β *and simple root* α_i.

Proof. Let us consider the inner products (α, α_i). If all of them are non-positive, then, by Lemma 7.17, $\{\alpha, \alpha_1, \ldots, \alpha_r\}$ are linearly independent, which is impossible since $\{\alpha_i\}$ is a basis. Thus, there exists i such that $(\alpha, \alpha_i) > 0$. Thus, $(\alpha, -\alpha_i) < 0$. By Lemma 7.11, this implies that $\beta = \alpha - \alpha_i$ is a root, so $\alpha = \beta + \alpha_i$. We leave it to the reader to check that β must be a positive root. This completes the proof of the lemma. $\qquad\square$

By Theorem 6.44, under the assumption of the lemma we have $\mathfrak{g}_\alpha = [\mathfrak{g}_\beta, e_i]$. Using induction in height $\mathrm{ht}(\alpha)$ (see equation (7.8)), it is now easy to show that e_i generate $\mathfrak{n}_+$. Similar argument shows that f_i generate $\mathfrak{n}_-$.

(3) Relations (7.27), (7.28) are an immediate corollary of the definition of Cartan subalgebra and root subspace. Commutation relation $[e_i, f_i] = h_i$ is part of Lemma 6.42 (about $\mathfrak{sl}(2, \mathbb{C})$-triple determined by a root). Commutation relation $[e_i, f_j] = 0$ for $i \neq j$ follows from the fact that $[e_i, f_j] \in \mathfrak{g}_{\alpha_i - \alpha_j}$. But $\alpha_i - \alpha_j$ is not a root (it can not be a positive root because the coefficient of α_j is negative, and it can not be a negative root because the coefficient of α_i is positive). Thus, $[e_i, f_j] = 0$.

To prove relations (7.31), consider the subspace $\bigoplus_{k \in \mathbb{Z}} \mathfrak{g}_{\alpha_j + k\alpha_i} \subset \mathfrak{g}$ as a module over $\mathfrak{sl}(2, \mathbb{C})$ triple generated by e_i, f_i, h_i. Since $\mathrm{ad}\, e_i.f_j = 0, f_j$ is a highest-weight vector; by (7.28), its weight is equal to $-a_{ij}$. Results of Section 4.8 about representation theory of $\mathfrak{sl}(2, \mathbb{C})$, imply that if v is a vector of weight λ in a finite-dimensional representation, with $e.v = 0$, then $f^{\lambda+1}.v = 0$. Applying it to f_j, we get (7.31). Equality (7.30) is proved similarly.

This completes the proof of Theorem 7.52. $\qquad\square$

A natural question is whether $(7.27) - (7.31)$ is a full set of defining relations for $\mathfrak{g}$. The answer is given by the following theorem.

Theorem 7.54. *Let R be a reduced irreducible root system, with a polarization $R = R_+ \sqcup R_-$ and system of simple roots $\Pi = \{\alpha_1, \ldots, \alpha_r\}$. Let $\mathfrak{g}(R)$ be the complex Lie algebra with generators $e_i, f_i, h_i, i = 1 \ldots, r$ and relations $(7.27) - (7.31)$. Then $\mathfrak{g}$ is a finite-dimensional semisimple Lie algebra with root system R.*

The proof of this theorem is not given here; interested reader can find it in [47], [22], or [24]. We note only that it is highly non-trivial that $\mathfrak{g}(R)$ is finite-dimensional (in fact, this is the key step of the proof), which in turn is based on the use of the Weyl group.

Corollary 7.55.

(1) *If $\mathfrak{g}$ is a semisimple Lie algebra with root system R, then there is a natural isomorphism $\mathfrak{g} \simeq \mathfrak{g}(R)$.*

(2) *There is a natural bijection between the set of isomorphism classes of reduced root systems and the set of isomorphism classes of finite-dimensional complex semisimple Lie algebras. The Lie algebra is simple iff the root system is irreducible.*

Combining this corollary with the classification given in Theorem 7.49, we get the following celebrated result.

Theorem 7.56. *Simple finite-dimensional complex Lie algebras are classified by Dynkin diagrams $A_n \dots G_2$ listed in Theorem 7.49.*

It is common to refer to the simple Lie algebra corresponding to the Dynkin diagram, say, E_6, as "simple Lie algebra of type E_6".

It is possible to give an explicit construction of the simple Lie algebra corresponding to each of the Dynkin diagrams of Theorem 7.49. For example, Lie algebra of type A_n is nothing but $\mathfrak{sl}(n + 1, \mathbb{C})$. Series B_n, C_n, D_n correspond to classical Lie algebras $\mathfrak{so}$ and $\mathfrak{sp}$. These root systems and Lie algebras are described in detail in Appendix A. Construction of exceptional Lie algebras, of types E_6, E_7, E_8, F_4, G_2, can be found in [3] or in [24].

7.10. Proof of the classification theorem in simply-laced case

In this section, we give a proof of the classification theorem for Dynkin diagrams (Theorem 7.49) in a special case when the diagram is simply-laced, i.e. contains no multiple edges. This section can be skipped at the first reading.

Let D be a connected simply-laced Dynkin diagram, with the set of vertices I. Then all roots α_i have the same length; without loss of generality, we can assume that $(\alpha_i, \alpha_i) = 2$. Then the Cartan matrix is given by $a_{ij} = (\alpha_i, \alpha_j)$. In particular, this implies the following important rule.

$$\text{For any } J \subset I, \text{ the matrix } (a_{ij})_{i,j \in J} \text{ is positive definite} \qquad (7.32)$$

We can now prove the classification theorem. For the reader's convenience, the proof is broken into several steps. In these arguments, a "subdiagram" is a diagram obtained by taking a subset of vertices of the original Dynkin diagram and all edges between them.

Step 1. D contains no cycles.

Indeed, otherwise D contains a subdiagram which is a cycle. But for such a subdiagram, bilinear form defined by the Cartan matrix is not positive definite: explicit computation shows that vector $\sum_{j \in J} \alpha_j$ is in the kernel of this form.

Step 2. Each vertex is connected to at most three others.

Indeed, otherwise D would contain the subdiagram shown in Figure 7.6. For such a subdiagram, however, the bilinear form defined by the Cartan matrix is not positive definite: vector $2\alpha + (\beta_1 + \beta_2 + \beta_3 + \beta_4)$ is in the kernel of this form.

Step 3. D contains at most one branching point (i.e., a vertex of valency more than 2).

Indeed, otherwise D contains the subdiagram shown in Figure 7.7.

Let $\alpha = \alpha_1 + \cdots + \cdots \alpha_n$. It is easy to see that $\alpha, \beta_1, \beta_2, \beta_3, \beta_4$ are linearly independent and thus the matrix of their inner products must be positive definite. On the other hand, explicit computation shows that $(\alpha, \alpha) = 2, (\alpha, \beta_i) = -1$, so the inner products between these vectors are given by the same matrix as in Step 2 and which, as was shown above, is not positive definite.

Combining the three steps, we see that D must be either a chain, i.e. a diagram of type A_n, or a "star" diagram with three branches as shown in Figure 7.8. Denote by p, q, r the lengths of these branches (including the central vertex); for example, for diagram E_7 the lengths would be $2, 3, 4$. Denote the roots corresponding to the branches of the diagram by $\beta_1, \ldots, \beta_{p-1}, \gamma_1, \ldots, \gamma_{q-1}, \delta_1, \ldots, \delta_{r-1}$ as shown in Figure 7.8.

Let α be the central vertex and let $\beta = \sum_{i=1}^{p-1} i\beta_i, \gamma = \sum_{i=1}^{q-1} i\gamma_i, \delta = \sum_{i=1}^{r-1} i\delta_i$. Then β, γ, δ are orthogonal, and vectors $\alpha, \beta, \gamma, \delta$ are linearly

Figure 7.6

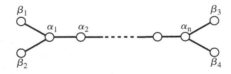

Figure 7.7

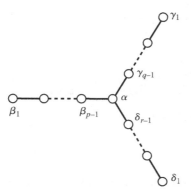

Figure 7.8

independent. Thus,

$$\left(\alpha, \tfrac{\beta}{|\beta|}\right)^2 + \left(\alpha, \tfrac{\gamma}{|\gamma|}\right)^2 + \left(\alpha, \tfrac{\delta}{|\delta|}\right)^2 < |\alpha|^2$$

or

$$\frac{(\alpha, \beta)^2}{(\beta, \beta)} + \frac{(\alpha, \gamma)^2}{(\gamma, \gamma)} + \frac{(\alpha, \delta)^2}{(\delta, \delta)} < 2.$$

Explicit computation shows that $(\beta, \beta) = p(p-1)$, and $(\alpha, \beta) = -(p-1)$, and similarly for γ, δ, so we get

$$\frac{p-1}{p} + \frac{q-1}{q} + \frac{r-1}{r} < 2$$

or

$$\frac{1}{p} + \frac{1}{q} + \frac{1}{r} > 1. \tag{7.33}$$

Since by definition $p, q, r \geq 2$, elementary analysis shows that up to the order, the only solutions are $(2, 2, n), n \geq 2$, $(2, 3, 3), (2, 3, 4), (2, 3, 5)$, which correspond to Dynkin diagrams D_{n+2}, E_6, E_7, E_8 respectively.

Thus we have shown that any simply-laced Dynkin diagram must be iso-morphic to one of A_n, D_n, E_6, E_7, E_8. It can be shown by explicitly constructing the corresponding root systems that each of these diagrams does appear as a Dynkin digram of a reduced root system. For D_n, such a construction is given in Appendix A; for E series, such a construction can be found in [47], [24] or [3]. This completes the proof of the classification theorem for simply-laced Dynkin diagrams.

The proof for the non-simply-laced case is quite similar but requires a couple of extra steps. This proof can be found in [22], [24], or [3] and will not be repeated here.

7.11. Exercises

7.1. Let $R \subset \mathbb{R}^n$ be given by

$$R = \{\pm e_i, \pm 2e_i \mid 1 \le i \le n\} \cup \{\pm e_i \pm e_j \mid 1 \le i,j \le n, i \ne j\},$$

where e_i is the standard basis in $\mathbb{R}^n$. Show that R is a non-reduced root system. (This root system is usually denoted BC_n.)

7.2. (1) Let $R \subset E$ be a root system. Show that the set

$$R^\vee = \{\alpha^\vee \mid \alpha \in R\} \subset E^*,$$

where $\alpha^\vee \in E^*$ is the coroot defined by (7.4), is also a root system. It is usually called the *dual root system* of R.

(2) Let $\Pi = \{\alpha_1, \ldots, \alpha_r\} \subset R$ be the set of simple roots. Show that the set $\Pi^\vee = \{\alpha_1^\vee, \ldots, \alpha_r^\vee\} \subset R^\vee$ is the set of simple roots of $R^\vee$. [Note: this is not completely trivial, as $\alpha \mapsto \alpha^\vee$ is not a linear map. Try using equation (7.17).]

7.3. Prove Lemma 7.17. (Hint: any linear dependence can be written in the form

$$\sum_{i \in I} c_i v_i = \sum_{j \in J} c_j v_j,$$

where $I \cap J = \varnothing, c_i, c_j \ge 0$. Show that if one denotes $v = \sum_{i \in I} c_i v_i$, then $(v, v) \le 0$.)

7.4. Show that $|P/Q| = |\det A|$, where A is the Cartan matrix: $a_{ij} = \langle \alpha_i^\vee, \alpha_j \rangle$.

7.5. Compute explicitly the group P/Q for root systems A_n, D_n.

7.6. Complete the gap in the proof of Theorem 7.37. Namely, assume that $w = s_{i_1} \ldots s_{i_l}$. Let $\beta_k = s_{i_1} \ldots s_{i_{k-1}}(\alpha_{i_k})$. Show that if we have $\beta_k = \pm \beta_j$ for some $j < k$ (thus, the path constructed in the proof of Theorem 7.37 crosses the hyperplane L_{β_j} twice), then $w = s_{i_1} \ldots \widehat{s_{i_j}} \ldots \widehat{s_{i_k}} \ldots s_{i_l}$ (hat means that the corresponding factor should be skipped) and thus, the original expression was not reduced.

7.7. Let $w = s_{i_1} \ldots s_{i_l}$ be a reduced expression. Show that then

$$\{\alpha \in R_+ \mid w(\alpha) \in R_-\} = \{\beta_1, \ldots, \beta_l\}$$

where $\beta_k = s_{i_1} \ldots s_{i_{k-1}}(\alpha_{i_k})$ (cf. proof of Lemma 7.31).

7.8. Let $\overline{C}_+$ be the closure of the positive Weyl chamber, and $\lambda \in \overline{C}_+, w \in W$ be such that $w(\lambda) \in \overline{C}_+$.
  (1) Show that $\lambda \in \overline{C}_+ \cap w^{-1}(\overline{C}_+)$.
  (2) Let $L_\alpha \subset E$ be a root hyperplane which separates C_+ and $w^{-1}C_+$. Show that then $\lambda \in L_\alpha$.
  (3) Show that $w(\lambda) = \lambda$.

Deduce from this that every W-orbit in E contains a unique element from $\overline{C}_+$.

7.9. Let $w_0 \in W$ be the longest element in the Weyl group W as defined in Lemma 7.39. Show that then for any $w \in W$, we have $l(ww_0) = l(w_0w) = l(w_0) - l(w)$.

7.10. Let $W = S_n$ be the Weyl group of root system A_{n-1}. Show that the longest element $w_0 \in W$ is the permutation $w_0 : i \mapsto n + 1 - i$.

7.11.

  (1) Let R be a reduced root system of rank 2, with simple roots α_1, α_2. Show that the longest element in the corresponding Weyl group is

$$w_0 = s_1 s_2 s_1 \cdots = s_2 s_1 s_2 \ldots \quad (m \text{ factors in each of the products})$$

where m depends on the angle φ between α_1, α_2: $\varphi = \pi - \frac{\pi}{m}$ (so $m = 2$ for $A_1 \cup A_1, m = 3$ for $A_2, m = 4$ for $B_2, m = 6$ for G_2). If you can not think of any other proof, give a case-by-case proof.
  (2) Show that the following relations hold in W (these are called *Coxeter relations*):

$$\begin{aligned} s_i^2 &= 1 \\ (s_i s_j)^{m_{ij}} &= 1, \end{aligned} \tag{7.34}$$

where m_{ij} is determined by the angle between α_i, α_j in the same way as in the previous part. (It can be shown that the Coxeter relations is a defining set of relations for the Weyl group: W could be defined as the group generated by elements s_i subject to Coxeter relations. A proof of this fact can be found in [23] or in [3].)

7.12. Let $\varphi \colon R_1 \xrightarrow{\sim} R_2$ be an isomorphism between irreducible root systems. Show that then φ is a composition of an isometry and a scalar operator: $(\varphi(v), \varphi(w)) = c(v, w)$ for any $v, w \in E_1$.

7.13. (1) Let $\mathfrak{n}_\pm$ be subalgebras in a semisimple complex Lie algebra defined by (7.25). Show that $\mathfrak{n}_\pm$ are nilpotent.

(2) Let $\mathfrak{b} = \mathfrak{n}_+ \oplus \mathfrak{h}$. Show that $\mathfrak{b}$ is solvable.

7.14. (1) Show that if two vertices in a Dynkin diagram are connected by a single edge, then the corresponding simple roots are in the same W-orbit.

(2) Show that for a reduced irreducible root system, the Weyl group acts transitively on the set of all roots of the same length.

7.15. Let $R \subset E$ be an irreducible root system. Show that then E is an irreducible representation of the Weyl group W.

7.16. Let G be a connected complex Lie group such that $\mathfrak{g} = \mathrm{Lie}(G)$ is semisimple. Fix a root decomposition of $\mathfrak{g}$.

(1) Choose $\alpha \in R$ and let $i_\alpha \colon \mathfrak{sl}(2, \mathbb{C}) \to \mathfrak{g}$ be the embedding constructed in Lemma 6.42; by Theorem 3.41, this embedding can be lifted to a morphism $i_\alpha \colon \mathrm{SL}(2, \mathbb{C}) \to G$.

Let

$$S_\alpha = i_\alpha \begin{pmatrix} 0 & -1 \\ 1 & 0 \end{pmatrix} = \exp\left(\frac{\pi}{2}(f_\alpha - e_\alpha)\right) \in G$$

(cf. Exercise 3.18). Show that $\mathrm{Ad}\, S_\alpha(h_\alpha) = -h_\alpha$ and that $\mathrm{Ad}\, S_\alpha(h) = h$ if $h \in \mathfrak{h}$, $\langle h, \alpha \rangle = 0$. Deduce from this that the action of S_α on $\mathfrak{g}^*$ preserves $\mathfrak{h}^*$ and that restriction of $\mathrm{Ad}\, S_\alpha$ to $\mathfrak{h}^*$ coincides with the reflection s_α.

(2) Show that the Weyl group W acts on $\mathfrak{h}^*$ by inner automorphisms: for any $w \in W$, there exists an element $\tilde{w} \in G$ such that $\mathrm{Ad}\, \tilde{w}|_{\mathfrak{h}^*} = w$. [Note, however, that in general, $\widetilde{w_1 w_2} \neq \tilde{w}_1 \tilde{w}_2$.]

7.17. Let

$$R = \{\pm e_i \pm e_j, i \neq j\} \cup \left\{\frac{1}{2} \sum_{i=1}^{8} \pm e_i\right\} \subset \mathbb{R}^8$$

(in the first set, signs are chosen independently; in the second, even number of signs should be pluses). Prove that R is a reduced root system with Dynkin diagram E_8.

8

Representations of semisimple Lie algebras

In this chapter, we study representations of complex semisimple Lie algebras. Recall that by results of Section 6.3, every finite-dimensional representation is completely reducible and thus can be written in the form $V = \bigoplus n_i V_i$, where V_i are irreducible representations and $n_i \in \mathbb{Z}_+$ are the multiplicities. Thus, the study of representations reduces to classification of irreducible representations and finding a way to determine, for a given representation V, the multiplicities n_i. Both of these questions have a complete answer, which will be given below.

Throughout this chapter, $\mathfrak{g}$ is a complex finite-dimensional semisimple Lie algebra. We fix a choice of a Cartan subalgebra and thus the root decomposition $\mathfrak{g} = \mathfrak{h} \oplus \bigoplus_R \mathfrak{g}_\alpha$ (see Section 6.6). We will freely use notation from Chapter 7; in particular, we denote by α_i, $i = 1 \ldots r$, simple roots, and by $s_i \in W$ corresponding simple reflections. We will also choose a non-degenerate invariant symmetric bilinear form $(\,,\,)$ on $\mathfrak{g}$.

All representations considered in this chapter are complex and unless specified otherwise, finite-dimensional.

8.1. Weight decomposition and characters

As in the study of representations of $\mathfrak{sl}(2, \mathbb{C})$ (see Section 4.8), the key to the study of representations of $\mathfrak{g}$ is decomposing the representation into the eigenspaces for the Cartan subalgebra.

Definition 8.1. Let V be a representation of $\mathfrak{g}$. A vector $v \in V$ is called a vector of weight $\lambda \in \mathfrak{h}^*$ if, for any $h \in \mathfrak{h}$, one has $hv = \langle \lambda, h \rangle v$. The space of all vectors of weight λ is called the *weight space* and denoted $V[\lambda]$:

$$V[\lambda] = \{v \in V \mid hv = \langle \lambda, h \rangle v \ \forall h \in \mathfrak{h}\}. \tag{8.1}$$

If $V[\lambda] \neq \{0\}$, then λ is called a weight of V. The set of all weights of V is denoted by $P(V)$:

$$P(V) = \{\lambda \in \mathfrak{h}^* \mid V[\lambda] \neq \{0\} \}. \tag{8.2}$$

Note that it easily follows from standard linear algebra results that vectors of different weights are linearly independent. This, in particular, implies that $P(V)$ is finite for a finite-dimensional representation.

Theorem 8.2. *Every finite-dimensional representation of* $\mathfrak{g}$ *admits a weight decomposition:*

$$V = \bigoplus_{\lambda \in P(V)} V[\lambda]. \tag{8.3}$$

Moreover, all weights of V are integral: $P(V) \subset P$, where P is the weight lattice defined in Section 7.5

Proof. Let $\alpha \in R$ be a root. Consider the corresponding $\mathfrak{sl}(2, \mathbb{C})$ subalgebra in $\mathfrak{g}$ generated by $e_\alpha, f_\alpha, h_\alpha$ as in Lemma 6.42. Considering V as a module over this $\mathfrak{sl}(2, \mathbb{C})$ and using the results of Section 4.8, we see that h_α is a diagonalizable operator in V. Since elements $h_\alpha, \alpha \in R$, span $\mathfrak{h}$, and the sum of the commuting diagonalizable operators is diagonalizable, we see that any $h \in \mathfrak{h}$ is diagonalizable. Since $\mathfrak{h}$ is commutative, all of them can be diagonalized simultaneously, which gives the weight decomposition.

Since weights of $\mathfrak{sl}(2, \mathbb{C})$ must be integer, we see that for any weight λ of V, we must have $\langle \lambda, h_\alpha \rangle \in \mathbb{Z}$, which by definition implies that $\lambda \in P$. $\qquad \square$

As in the $\mathfrak{sl}(2, \mathbb{C})$ case, this weight decomposition agrees with the root decomposition of $\mathfrak{g}$.

Lemma 8.3. *If $x \in \mathfrak{g}_\alpha$, then $x.V[\lambda] \subset V[\lambda + \alpha]$.*

Proof of this lemma is almost identical to the proof in $\mathfrak{sl}(2, \mathbb{C})$ case (see Lemma 4.55). Details are left to the reader.

For many practical applications it is important to know the dimensions of the weight subspaces $V[\lambda]$. To describe them, it is convenient to introduce the formal generating series for these dimensions as follows.

Let $\mathbb{C}[P]$ be the algebra generated by formal expressions e^λ, $\lambda \in P$, subject to the following relations:

$$\begin{aligned} e^\lambda e^\mu &= e^{\lambda + \mu} \\ e^0 &= 1 \end{aligned} \tag{8.4}$$

Algebra $\mathbb{C}[P]$ can also be described as the algebra of polynomial complex-valued functions on the torus $T = \mathfrak{h}/2\pi i Q^\vee$, where $Q^\vee$ is the coroot lattice defined in Section 7.5, by letting

$$e^\lambda(t) = e^{\langle t, \lambda \rangle}, \qquad \lambda \in P, \quad t \in \mathfrak{h}/2\pi i Q^\vee \tag{8.5}$$

which explains the notation. It is easy to show that algebra $\mathbb{C}[P]$ is isomorphic to the algebra of Laurent polynomials in $r = \text{rank } \mathfrak{g}$ variables (see Exercise 8.3); thus, we will commonly refer to elements of $\mathbb{C}[P]$ as polynomials.

Definition 8.4. Let V be a finite-dimensional representation of $\mathfrak{g}$. We define its character $\text{ch}(V) \in \mathbb{C}[P]$ by

$$\text{ch}(V) = \sum (\dim V[\lambda]) e^\lambda.$$

Remark 8.5. Note that the word "character" had already been used before, in relation with group representations (see Definition 4.43). In fact, these two definitions are closely related: any finite-dimensional representation of $\mathfrak{g}$ can also be considered as a representation of the corresponding simply-connected complex Lie group G. In particular, every $t \in \mathfrak{h}$ gives an element $\exp(t) \in G$. Then it follows from the definition that if we consider elements of $\mathbb{C}[P]$ as functions on $T = \mathfrak{h}/2\pi i Q^\vee$ as defined in (8.5), then

$$\text{ch}(V)(t) = \text{tr}_V(\exp(t))$$

which establishes the relation with Definition 4.43.

Example 8.6. Let $\mathfrak{g} = \mathfrak{sl}(2, \mathbb{C})$. Then $P = \mathbb{Z}\frac{\alpha}{2}$, so $\mathbb{C}[P]$ is generated by $e^{n\alpha/2}$, $n \in \mathbb{Z}$. Denoting $e^{\alpha/2} = x$, we see that $\mathbb{C}[P] = \mathbb{C}[x, x^{-1}]$. By Theorem 4.59, the character of irreducible representation V_n is given by

$$\text{ch}(V_n) = x^n + x^{n-2} + x^{n-4} + \cdots + x^{-n} = \frac{x^{n+1} - x^{-n-1}}{x - x^{-1}}.$$

The following lemma lists some basic properties of characters.

Lemma 8.7.

(1) $\text{ch}(\mathbb{C}) = 1$
(2) $\text{ch}(V_1 \oplus V_2) = \text{ch}(V_1) + \text{ch}(V_2)$
(3) $\text{ch}(V_1 \otimes V_2) = \text{ch}(V_1) \text{ch}(V_2)$
(4) $\text{ch}(V^*) = \overline{\text{ch}(V)}$, *where* $\overline{}$ *is defined by*

$$\overline{e^\lambda} = e^{-\lambda}.$$

Proof of all of these facts is left to the reader as an easy exercise.

In the example of $\mathfrak{sl}(2, \mathbb{C})$ one notices that the characters are symmetric with respect to the action of the Weyl group (which in this case acts by $x \mapsto x^{-1}$). It turns out that a similar result holds in general.

Theorem 8.8. *If V is a finite-dimensional representation of* $\mathfrak{g}$*, then the set of weights and dimensions of weight subspaces are Weyl group invariant: for any* $w \in W$, $\dim V[\lambda] = \dim V[w(\lambda)]$. *Equivalently,*

$$w(\mathrm{ch}(V)) = \mathrm{ch}(V),$$

where the action of W on $\mathbb{C}[P]$ *is defined by*

$$w(\mathrm{e}^{\lambda}) = \mathrm{e}^{w(\lambda)}.$$

Proof. Since W is generated by simple reflections s_i, it suffices to prove this theorem for $w = s_i$. Let $\langle \lambda, \alpha_i^{\vee} \rangle = n \geq 0$; then it follows from the representation theory of $\mathfrak{sl}(2, \mathbb{C})$ (Theorem 4.60) that operators $f_i^n \colon V[\lambda] \to V[\lambda - n\alpha_i]$ and $e_i^n \colon V[\lambda - n\alpha_i] \to V[\lambda]$ are isomorphisms (in fact, up to a constant, they are mutually inverse) and thus $\dim V[\lambda] = \dim V[\lambda - n\alpha_i]$. Since $\lambda - n\alpha_i = \lambda - \langle \lambda, \alpha_i^{\vee} \rangle \alpha_i = s_i(\lambda)$, this shows that dimensions of the weight subspaces are invariant under s_i. $\qquad\square$

Later we will show that characters of irreducible finite-dimensional representations form a basis of the subalgebra of W-invariants $\mathbb{C}[P]^W \subset \mathbb{C}[P]$ (see Theorem 8.41).

Figure 8.1 shows an example of the set of weights of a representation of $\mathfrak{sl}(3, \mathbb{C})$.

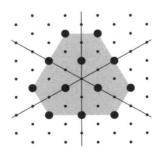

Figure 8.1 Set of weights of a representation V of Lie algebra $\mathfrak{sl}(3, \mathbb{C})$. Large dots are the weights of V, small dots are all weights $\lambda \in P$; shaded area is the convex hull of $P(V)$. The figure also shows the root hyperplanes.

8.2. Highest weight representations and Verma modules

To study irreducible representations, we introduce a class of representations that
are generated by a single vector. As we will later show, all finite-dimensional
irreducible representations fall into this class. However, it turns out that to study
finite-dimensional representations, we need to consider infinite-dimensional
representations as an auxiliary tool.

Recall (see Theorem 7.52) that choice of polarization of the root system gives
the following decomposition for the Lie algebra $\mathfrak{g}$:

$$\mathfrak{g} = \mathfrak{n}_- \oplus \mathfrak{h} \oplus \mathfrak{n}_+, \qquad \mathfrak{n}_\pm = \bigoplus_{\alpha \in R_\pm} \mathfrak{g}_\alpha.$$

Definition 8.9. A non-zero representation V (not necessarily finite-
dimensional) of $\mathfrak{g}$ is called a *highest weight representation* if it is generated
by a vector $v \in V[\lambda]$ such that $x.v = 0$ for all $x \in \mathfrak{n}_+$. In this case, v is called
the *highest weight vector*, and λ is the highest weight of V.

The importance of such representations is explained by the following
theorem.

Theorem 8.10. *Every irreducible finite-dimensional representation of $\mathfrak{g}$ is a
highest weight representation.*

Proof. Let $P(V)$ be the set of weights of V. Let $\lambda \in P(V)$ be such that for all
$\alpha \in R_+$, $\lambda + \alpha \notin P(V)$. Such a λ exists: for example, we can take $h \in \mathfrak{h}$ such
that $\langle h, \alpha \rangle > 0$ for all $\alpha \in R_+$, and then consider $\lambda \in P(V)$ such that $\langle h, \lambda \rangle$ is
maximal possible.

Now let $v \in V[\lambda]$ be a non-zero vector. Since $\lambda + \alpha \notin P(V)$, we have
$e_\alpha v = 0$ for any $\alpha \in R_+$. Consider the subrepresentation $V' \subset V$ generated by
v. By definition, V' is a highest weight representation. On the other hand, since
V is irreducible, one has $V' = V$. $\qquad\square$

Note that there can be many non-isomorphic highest weight representations
with the same highest weight. However, in any highest weight representation
with highest weight vector $v_\lambda \in V[\lambda]$, the following conditions hold:

$$\begin{aligned}
hv_\lambda &= \langle h, \lambda \rangle v_\lambda \quad \forall h \in \mathfrak{h} \\
xv_\lambda &= 0 \quad \forall x \in \mathfrak{n}_+.
\end{aligned} \tag{8.6}$$

Let us define the universal highest weight representation M_λ as the representation generated by a vector v_λ satisfying conditions (8.6) and no other relations.

More formally, define

$$M_\lambda = U\mathfrak{g}/I_\lambda, \tag{8.7}$$

where I_λ is the left ideal in $U\mathfrak{g}$ generated by vectors $e \in \mathfrak{n}_+$ and $(h - \langle h, \lambda \rangle), h \in \mathfrak{h}$. This module is called the *Verma module* and plays an important role in representation theory.

Alternatively, Verma modules can be defined as follows. Define the *Borel subalgebra* $\mathfrak{b}$ by

$$\mathfrak{b} = \mathfrak{h} \oplus \mathfrak{n}_+. \tag{8.8}$$

Formulas (8.6) define a one-dimensional representation of $\mathfrak{b}$ which we will denote by $\mathbb{C}_\lambda$. Then Verma module M_λ can be defined by

$$M_\lambda = U\mathfrak{g} \otimes_{U\mathfrak{b}} \mathbb{C}_\lambda. \tag{8.9}$$

Remark 8.11. Readers familiar with the notion of induced representation will recognize that M_λ can be naturally described as an induced representation: $M_\lambda = \mathrm{Ind}_{U\mathfrak{b}}^{U\mathfrak{g}} \mathbb{C}_\lambda$.

Example 8.12. Let $\mathfrak{g} = \mathfrak{sl}(2, \mathbb{C})$ and identify $\mathfrak{h}^* \simeq \mathbb{C}$ by $\lambda \mapsto \langle h, \lambda \rangle$, so that $\alpha \mapsto 2$. Then Verma module $M_\lambda, \lambda \in \mathbb{C}$, is the module described in Lemma 4.58.

The following lemma shows that Verma modules are indeed universal in a suitable sense.

Lemma 8.13. *If V is a highest weight representation with highest weight λ, then*

$$V \simeq M_\lambda/W,$$

for some submodule $W \subset M_\lambda$.

Thus, the study of highest weight representations essentially reduces to the study of submodules in Verma modules.

Theorem 8.14. *Let $\lambda \in \mathfrak{h}^*$ and let M_λ be the Verma module with highest weight λ.*

(1) *Every vector $v \in M_\lambda$ can be uniquely written in the form $v = uv_\lambda, u \in U\mathfrak{n}_-$. In other words, the map*

$$U\mathfrak{n}_- \to M_\lambda$$

$$u \mapsto uv_\lambda,$$

is an isomorphism of vector spaces.

(2) M_λ admits a weight decomposition: $M_\lambda = \bigoplus_\mu M_\lambda[\mu]$, with finite-dimensional weight spaces. The set of weights of M_λ is

$$P(M_\lambda) = \lambda - Q_+, \qquad Q_+ = \left\{ \sum n_i \alpha_i, n_i \in \mathbb{Z}_+ \right\} \qquad (8.10)$$

(3) $\dim M_\lambda[\lambda] = 1$.

Proof. By a corollary of the PBW theorem (Corollary 5.14), since $\mathfrak{g} = \mathfrak{n}_- \oplus \mathfrak{b}$, $U\mathfrak{g} \simeq U\mathfrak{n}_- \otimes U\mathfrak{b}$ as an $U\mathfrak{n}_-$-module. Therefore, using (8.9), we have

$$M_\lambda = U\mathfrak{g} \otimes_{U\mathfrak{b}} \mathbb{C}_\lambda = U\mathfrak{n}_- \otimes U\mathfrak{b} \otimes_{U\mathfrak{b}} \mathbb{C}_\lambda = U\mathfrak{n}_- \otimes \mathbb{C}_\lambda,$$

which proves (1). Parts (2) and (3) immediately follow from (1). $\qquad \square$

Figure 8.2 shows set of weights of a Verma module over $\mathfrak{sl}(3, \mathbb{C})$.

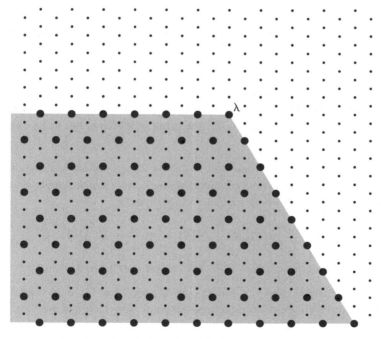

Figure 8.2 Set of weights of a Verma module M_λ for Lie algebra $\mathfrak{sl}(3, \mathbb{C})$. Large dots are the weights of M_λ, small dots are all weights $\mu \in P$; shaded area is the convex hull of $P(M_\lambda)$.

Since every highest weight representation is a quotient of a Verma module, the above theorem can be generalized to an arbitrary highest weight representation. For future convenience, introduce relations $\prec, \preceq$ on $\mathfrak{h}^*$ by

$$\lambda \preceq \mu \text{ iff } \mu - \lambda \in Q_+,$$
$$\lambda \prec \mu \iff \lambda \preceq \mu, \lambda \neq \mu,$$

(8.11)

where Q_+ is defined by (8.10). It is easy to see that $\preceq$ is a partial order on $\mathfrak{h}^*$.

Theorem 8.15. *Let V be a highest weight representation with highest weight λ (not necessarily finite-dimensional).*

(1) *Every vector $v \in V$ can be written in the form $v = uv_\lambda, u' \in U\mathfrak{n}_-$. In other words, the map*

$$U\mathfrak{n}_- \to V$$

$$u \mapsto uv_\lambda$$

is surjective.

(2) *V admits a weight decomposition: $V = \bigoplus_{\mu \preceq \lambda} V[\mu]$, with finite-dimensional weight subspaces.*

(3) *$\dim V[\lambda] = 1$.*

Proof. Part (1) immediately follows from the similar statement for Verma modules. Part (2) also follows from weight decomposition for Verma modules and the following linear algebra lemma, the proof of which is left as an exercise (see Exercise 8.1).

Lemma 8.16. *Let $\mathfrak{h}$ be a commutative finite-dimensional Lie algebra and M a module over $\mathfrak{h}$ (not necessarily finite-dimensional) which admits a weight decomposition with finite-dimensional weight spaces:*

$$M = \bigoplus M[\lambda], \qquad M[\lambda] = \{v \mid hv = \langle h, \lambda \rangle v\}.$$

Then any submodule, quotient of M also admits a weight decomposition.

To prove part (3), note that $\dim V[\lambda] \leq \dim M_\lambda[\lambda] = 1$. On the other hand, by definition of a highest weight representation, V does have a non-zero highest weight vector $v \in V[\lambda]$. $\qquad \square$

Corollary 8.17. *In any highest weight representation, there is a unique highest weight and unique up to a scalar highest weight vector.*

Proof. Indeed, if λ, μ are highest weights, then by Theorem 8.15, $\lambda \preceq \mu$ and $\mu \preceq \lambda$, which is impossible unless $\lambda = \mu$. □

8.3. Classification of irreducible finite-dimensional representations

Our next goal is to classify all irreducible finite-dimensional representations. Since by Theorem 8.10 every such representation is a highest weight representation, this question can be reformulated as follows: classify all highest weight representations which are finite-dimensional and irreducible.

The first step is the following easy result.

Theorem 8.18. *For any $\lambda \in \mathfrak{h}^*$, there exists a unique up to isomorphism irreducible highest weight representation with highest weight λ. This representation is denoted L_λ.*

Proof. All highest weight representations with highest weight λ are of the form M_λ/W for some $W \subset M_\lambda$. It is easy to see that M_λ/W is irreducible iff W is a maximal proper subrepresentation (that is, not properly contained in any other proper subrepresentation). Thus, it suffices to prove that M_λ has a unique maximal proper submodule.

Note that by Lemma 8.16, every proper submodule $W \subset M_\lambda$ admits a weight decomposition and $W[\lambda] = 0$ (otherwise, we would have $W[\lambda] = M_\lambda[\lambda]$, which would force $W = M_\lambda$). Let J_λ be the sum of all submodules $W \subset M_\lambda$ such that $W[\lambda] = 0$. Then $J_\lambda \subset M_\lambda$ is proper (because $J_\lambda[\lambda] = 0$). Since it contains every other proper submodule of M_λ, it is the unique maximal proper submodule of M_λ. Thus, $L_\lambda = M_\lambda/J_\lambda$ is the unique irreducible highest-weight module with highest weight λ. □

Example 8.19. For $\mathfrak{g} = \mathfrak{sl}(2, \mathbb{C})$, the results of Section 4.8 show that if $\lambda \in \mathbb{Z}_+$, then $L_\lambda = V_\lambda$ is the finite-dimensional irreducible module of dimension $\lambda + 1$, and $L_\lambda = M_\lambda$ for $\lambda \notin \mathbb{Z}_+$.

As we will later see, the situation is similar for other Lie algebras. In particular, for "generic" λ, the Verma module is irreducible, so $M_\lambda = L_\lambda$.

Since every irreducible finite-dimensional representation is a highest weight representation (Theorem 8.10), we get the following corollary.

Corollary 8.20. *Every irreducible finite-dimensional representation V must be isomorphic to one of L_λ.*

Thus, to classify all irreducible finite-dimensional representations of $\mathfrak{g}$, we need to find which of L_λ are finite-dimensional.

To give an answer to this question, we need to introduce some notation. Recall the weight lattice $P \subset \mathfrak{h}^*$ defined by (7.12).

Definition 8.21. A weight $\lambda \in \mathfrak{h}^*$ is called *dominant integral* if the following condition holds

$$\langle \lambda, \alpha^\vee \rangle \in \mathbb{Z}_+ \qquad \text{for all } \alpha \in R_+. \tag{8.12}$$

The set of all dominant integral weights is denoted by P_+.

It follows from results of Exercise 7.2 that condition (8.12) is equivalent to

$$\langle \lambda, \alpha_i^\vee \rangle \in \mathbb{Z}_+ \qquad \text{for all } \alpha_i \in \Pi. \tag{8.13}$$

Lemma 8.22.

(1) $P_+ = P \cap \overline{C}_+$, where $C_+ = \{\lambda \in \mathfrak{h}^* \mid \langle \lambda, \alpha_i^\vee \rangle > 0 \; \forall i\}$ *is the positive Weyl chamber and $\overline{C}_+$ is its closure.*
(2) *For any $\lambda \in P$, its Weyl group orbit $W\lambda$ contains exactly one element of P_+.*

Proof. The first part is immediate from the definitions. The second part follows from the fact that any W-orbit in $\mathfrak{h}_\mathbb{R}^*$ contains exactly one element from $\overline{C}_+$ (Exercise 7.8). □

Theorem 8.23. *Irreducible highest weight representation L_λ is finite-dimensional iff $\lambda \in P_+$.*

Before proving this theorem, note that together with Theorem 8.10 it immediately implies the following corollary.

Corollary 8.24. *For every $\lambda \in P_+$, representation L_λ is an irreducible finite-dimensional representation. These representations are pairwise non-isomorphic, and every irreducible finite-dimensional representation is isomorphic to one of them.*

Proof of Theorem 8.23. First, let us prove that if L_λ is finite-dimensional, then $\lambda \in P_+$. Indeed, let α_i be a simple root and let $\mathfrak{sl}(2, \mathbb{C})_i$ be the subalgebra in $\mathfrak{g}$ generated by $e_{\alpha_i}, h_{\alpha_i}, f_{\alpha_i}$ (see Lemma 6.42). Consider L_λ as $\mathfrak{sl}(2, \mathbb{C})_i$ module. Then v_λ satisfies relations $e_i v_\lambda = 0, h_i v_\lambda = \langle h_i, \lambda \rangle v_\lambda = \langle \alpha_i^\vee, \lambda \rangle v_\lambda$. It generates a highest weight $\mathfrak{sl}(2, \mathbb{C})_i$ submodule, which is finite-dimensional (since L_λ is finite-dimensional). By classification theorem for irreducible representation of $\mathfrak{sl}(2, \mathbb{C})$ (Theorem 4.59), this implies that the highest weight $\langle h_i, \lambda \rangle \in \mathbb{Z}_+$. Repeating the argument for each simple root, we see that $\lambda \in P_+$.

Now let us prove that if $\lambda \in P_+$, then L_λ is finite-dimensional. This is a more difficult result; we break the proof in several steps.

Step 1. Let $n_i = \langle \alpha_i^\vee, \lambda \rangle \in \mathbb{Z}_+$. Consider the vector

$$v_{s_i.\lambda} = f_i^{n_i+1} v_\lambda \in M_\lambda[s_i.\lambda], \tag{8.14}$$

where $v_\lambda \in M_\lambda$ is the highest-weight vector and

$$s_i.\lambda = \lambda - (n_i + 1)\alpha_i. \tag{8.15}$$

(we will give a more general definition later, see (8.20)). Then we have

$$e_j v_{s_i.\lambda} = 0 \quad \text{for all } i, j. \tag{8.16}$$

Indeed, for $i \neq j$ we have $[e_j, f_i] = 0$ (see equation (7.29)), so $e_j f_i^{n_i+1} v_\lambda = f_i^{n_i+1} e_j v_\lambda = 0$. For $i = j$, this follows from the results of Section 4.8: if v is a vector of weight n in a representation of $\mathfrak{sl}(2, \mathbb{C})$ such that $ev = 0$, then $ef^{n+1}v = 0$.

Step 2. Let $M_i \subset M_\lambda$ be the subrepresentation generated by vector $v_{s_i.\lambda}$. By (8.16), M_i is a highest weight representation. In particular, by Theorem 8.15 all weights μ of M_i must satisfy $\mu \preceq s_i.\lambda \prec \lambda$. Thus, λ is not a weight of M_i; therefore, each M_i is a proper submodule in M_λ.

Consider now the quotient

$$\tilde{L}_\lambda = M_\lambda / \sum M_i. \tag{8.17}$$

Since each M_i is a proper subrepresentation, so is $\sum M_i$ (see the proof of Theorem 8.18); thus, $\tilde{L}_\lambda$ is a non-zero highest weight representation.

Step 3. The key step of the proof is the following theorem.

Theorem 8.25. *Let $\lambda \in P_+$, and let $\tilde{L}_\lambda$ be defined by (8.17). Then $\tilde{L}_\lambda$ is finite-dimensional.*

The proof of this theorem is rather long. It is given in a separate section (Section 8.9) at the end of this chapter.

Now we can complete the proof of Theorem 8.23. Since L_λ is the quotient of M_λ by the maximal proper ideal, we see that L_λ is a quotient of $\tilde{L}_\lambda$. Since $\tilde{L}_\lambda$ is finite-dimensional, so is L_λ. This completes the proof. (Later we will show that in fact $\tilde{L}_\lambda = L_\lambda$; see Theorem 8.28.) $\qquad\square$

Note that for many practical applications, construction of the irreducible finite-dimensional representation given in this theorem is not very useful, as it gives such a representation as a quotient of an infinite-dimensional representation M_λ. However, for all classical algebras there also exist very explicit constructions of the irreducible finite-dimensional representations, which are usually not based on realizing L_λ as quotients of Verma modules. We will give an example of this for $\mathfrak{g} = \mathfrak{sl}(n, \mathbb{C})$ in Section 8.7

Example 8.26. Let $\mathfrak{g}$ be a simple Lie algebra. Consider the adjoint representation of $\mathfrak{g}$, which in this case is irreducible. Weights of this representation are exactly $\alpha \in R$ (with multiplicity 1) and 0 with multiplicity dim $\mathfrak{h} = r$.

By general theory above, this representation must have a highest weight θ, which can be defined by conditions $\theta \in R$, $\theta + \alpha \notin R \cup \{0\}$ for any $\alpha \in R_+$; from this, it is easy to see that $\theta \in R_+$. Usually, θ is called the *maximal root* of $\mathfrak{g}$. Another characterization can be found in Exercise 8.6.

In particular, for $\mathfrak{g} = \mathfrak{sl}(n, \mathbb{C})$, the maximal root θ is given by $\theta = e_1 - e_n$.

8.4. Bernstein–Gelfand–Gelfand resolution

In the previous section, we have shown that for $\lambda \in P_+$, the irreducible highest weight representation L_λ is finite-dimensional. Our next goal is to study the structure of these representations – in particular, to find the dimensions of weight subspaces.

Recall that in the proof of Theorem 8.23 we defined, for each $\lambda \in P_+$, a collection of submodules $M_i \subset M_\lambda$. We have shown that each of them is a highest weight module. In fact, we can make a more precise statement.

Lemma 8.27. *Let $v \in M_\lambda[\mu]$ be a vector such that $\mathfrak{n}_+ v = 0$ (such a vector is called a* singular vector*), and let $M' \subset M_\lambda$ be the submodule generated by v. Then M' is a Verma module with highest weight μ.*

Proof. Since $e_\alpha v = 0$, by definition M' is a highest weight representation with highest weight μ and thus is isomorphic to a quotient of the Verma module M_μ. To show that $M' = M_\mu$, it suffices to show that the map $U\mathfrak{n}_- \to M' : u \mapsto uv$ is injective.

Indeed, assume that $uv = 0$ for some $u \in U\mathfrak{n}_-$. On the other hand, by Theorem 8.14, we can write $v = u'v_\lambda$ for some $u' \in U\mathfrak{n}_-$. Thus, we get $uu'v_\lambda = 0$. By Theorem 8.14, this implies that $uu' = 0$ as element of $U\mathfrak{n}_-$. But by Corollary 5.15, $U\mathfrak{n}_-$ has no zero divisors.   □

This allows us to give a better description of the irreducible finite-dimensional modules L_λ, $\lambda \in P_+$.

Theorem 8.28. *Let* $\lambda \in P_+$. *As in the proof of Theorem 8.23, let*

$$v_{s_i.\lambda} = f_i^{n_i+1} v_\lambda \in M_\lambda, \qquad n_i = \langle \lambda, \alpha_i^\vee \rangle$$

and let $M_i \subset M_\lambda$ *be the submodule generated by* $v_{s_i.\lambda}$.

(1) *Each* M_i *is isomorphic to a Verma module:*

$$M_i \simeq M_{s_i.\lambda}, \qquad s_i.\lambda = \lambda - (n_i + 1)\alpha_i$$

(2)

$$L_\lambda = M_\lambda / \sum_{i=1}^r M_i.$$

Proof. Part (1) is an immediate corollary of (8.16) and Lemma 8.27. To prove part (2), let $\tilde{L}_\lambda = M_\lambda / \sum M_i$. As was shown in the proof of Theorem 8.23, $\tilde{L}_\lambda$ is finite-dimensional, and all weights μ of $\tilde{L}_\lambda$ satisfy $\mu \preceq \lambda$. By complete reducibility (Theorem 6.20), we can write $\tilde{L}_\lambda = \bigoplus_{\mu \preceq \lambda, \ \mu \in P+} n_\mu L_\mu$. Comparing the dimensions of subspace of weight λ, we see that $n_\lambda = 1$: $\tilde{L}_\lambda = L_\lambda \oplus \left(\bigoplus_{\mu \prec \lambda} n_\mu L_\mu \right)$. This shows that the highest weight vector v_λ of $\tilde{L}_\lambda$ is in L_λ; thus, the submodule it generates must be equal to L_λ. On the other hand, v_λ generates $\tilde{L}_\lambda$. $\qquad\square$

Example 8.29. For $\mathfrak{g} = \mathfrak{sl}(2, \mathbb{C})$, this theorem reduces to $L_\lambda = M_\lambda / M_{-\lambda-1}$, $\lambda \in \mathbb{Z}_+$, which had already been established in the proof of Theorem 4.59.

Theorem 8.28 provides some description of the structure of finite-dimensional irreducible representations L_λ. In particular, we can try to use it to find dimensions of the weight subspaces in L_λ: indeed, for M_λ dimensions of the weight subspaces can be easily found, since by Theorem 8.14, for $\beta \in Q_+$, we have $\dim M_\lambda[\lambda - \beta] = \dim U\mathfrak{n}_-[-\beta]$. The latter dimension can be easily found using the PBW theorem.

However, for Lie algebras other than $\mathfrak{sl}(2, \mathbb{C})$, Theorem 8.28 does not give full information about the structure of L_λ; in particular, it is not enough to find the dimensions of the weight subspaces. The problem is that the sum $\sum M_i$ is not a direct sum: these modules have a non-trivial intersection. Another way to say it is as follows: Theorem 8.28 describes L_λ as the cokernel of the map

$$\bigoplus M_{s_i.\lambda} \to M_\lambda \qquad\qquad (8.18)$$

but does not describe its kernel. Since $L_\lambda = M_\lambda / \sum M_i$, we can extend (8.18) to the following exact sequence of modules:

$$\bigoplus M_{s_i.\lambda} \to M_\lambda \to L_\lambda \to 0 \tag{8.19}$$

which, unfortunately, is not a short exact sequence: the first map is not injective.

It turns out, however, that (8.19) can be extended to a long exact sequence, which gives a resolution of L_λ. All terms in this resolution will be direct sums of Verma modules.

Define the shifted action of the Weyl group W on $\mathfrak{h}^*$ by

$$w.\lambda = w(\lambda + \rho) - \rho, \tag{8.20}$$

where $\rho = \frac{1}{2} \sum_{\alpha \in R_+} \alpha$ as in Lemma 7.36. Note that in particular, $s_i \rho = \rho - \alpha_i$ (see proof of Lemma 7.36), so this definition agrees with earlier definition (8.15).

Theorem 8.30. *Let $\lambda \in P_+$. Then there exists a long exact sequence*

$$0 \to M_{w_0.\lambda} \to \cdots \bigoplus_{w \in W, l(w)=k} M_{w.\lambda} \cdots \to \bigoplus_i M_{s_i.\lambda} \to M_\lambda \to L_\lambda \to 0$$

where $l(w)$ is the length of an element $w \in W$ as defined in Definition 7.34, and w_0 is the longest element of the Weyl group (see Lemma 7.39).

This is called the Bernstein–Gelfand–Gelfand (BGG) resolution of L_λ.

The proof of this theorem is rather hard and will not be given here. The interested reader can find it in the original paper [2].

Example 8.31. For $\mathfrak{g} = \mathfrak{sl}(2, \mathbb{C})$, $W = \{1, s\}$, and $s.\lambda = -\lambda - 2$ (if we identify $\mathfrak{h}^*$ with $\mathbb{C}$ as in Example 8.12), so the BGG resolution takes the form

$$0 \to M_{-\lambda-2} \to M_\lambda \to L_\lambda \to 0$$

(compare with the proof of Theorem 4.59).

Example 8.32. For $\mathfrak{g} = \mathfrak{sl}(3, \mathbb{C})$, $W = S_3$, and the BGG resolution takes the form

$$0 \to M_{s_1 s_2 s_1.\lambda} \to \left(M_{s_1 s_2.\lambda} \oplus M_{s_2 s_1.\lambda} \right) \to \left(M_{s_1.\lambda} \oplus M_{s_2.\lambda} \right)$$
$$\to M_\lambda \to L_\lambda \to 0.$$

BGG resolution leads to many extremely interesting and deep connections between representation theory, geometry, and combinatorics. For the curious reader we briefly list some of the exciting possibilities it opens; none of them will be used in the remainder of this book.

- For fixed $\lambda \in P_+$, inclusions of the modules of the form $M_{w.\lambda}$ define a partial order on W. This order is independent of λ (as long as $\lambda \in P_+$) and is called *Bruhat order* on W.

- BGG resolution and Bruhat order are closely related to the geometry of the flag variety $\mathcal{F} = G/B$, where G is the simply-connected complex Lie group with Lie algebra $\mathfrak{g}$ and B is the Borel subgroup, i.e. the subgroup corresponding to the Borel subalgebra $\mathfrak{b}$; in the simplest example of $G = GL(n, \mathbb{C})$, the flag variety $\mathcal{F}$ was described in Example 2.25. In particular, the flag variety admits a cell decomposition in which the cells are labeled by elements of W and the partial order defined by "cell C_1 is in the closure of cell C_2" coincides with the Bruhat order.

- One can also ask if it is possible to write a similar resolution in terms of Verma modules for L_λ when $\lambda \notin P_+$. It can be shown that for generic λ, $L_\lambda = M_\lambda$; however, there is a number of intermediate cases between generic λ and $\lambda \in P_+$. It turns out that there is indeed an analog of BGG resolution for any λ, but it is highly non-trivial; proper description of it, given by Kazhdan and Lusztig, requires introducing a new cohomology theory (intersection cohomology, or equivalently, cohomology of perverse sheaves) for singular varieties such as the closures of cells in flag variety. An introduction to this theory, with further references, can be found in [37].

8.5. Weyl character formula

Recall that for a finite-dimensional representation V we have defined its character $\mathrm{ch}(V)$ by $\mathrm{ch}(V) = \sum \dim V[\lambda] e^\lambda \in \mathbb{C}[P]$. In this section, we will give an explicit formula for characters of irreducible representations L_λ.

Before doing this, we will need to define characters for certain infinite-dimensional representations; however, they will be not in $\mathbb{C}[P]$ but in a certain completion of it. There are several possible completions; the one we will use is defined by

$$\widehat{\mathbb{C}[P]} = \{f = \sum_{\lambda \in P} c_\lambda e^\lambda \mid \mathrm{supp} f \subset \text{finite union of cones } (\lambda_i - Q_+)\}, \quad (8.21)$$

where $\mathrm{supp} f = \{\lambda \mid c_\lambda \neq 0\}$.

In particular, it follows from Theorem 8.15 that for any highest weight representation M with integral highest weight, the character $\mathrm{ch}(M)$ defined as in

Definition 8.4 is in $\widehat{\mathbb{C}[P]}$. The same holds for finite direct sums of highest weight modules.

In particular, it is very easy to compute characters of Verma modules.

Lemma 8.33. *For any $\lambda \in P$,*

$$\mathrm{ch}(M_\lambda) = \frac{e^\lambda}{\prod_{\alpha \in R_+}(1 - e^{-\alpha})},$$

where each factor $\frac{1}{1-e^{-\alpha}}$ should be understood as a formal series

$$\frac{1}{1 - e^{-\alpha}} = 1 + e^{-\alpha} + e^{-2\alpha} + \cdots$$

Proof. Since $u \mapsto uv_\lambda$ gives an isomorphism $U\mathfrak{n}_- \simeq M_\lambda$ (Theorem 8.14), we see that $\mathrm{ch}(M_\lambda) = e^\lambda\,\mathrm{ch}(U\mathfrak{n}_-)$ (note that $U\mathfrak{n}_-$ is not a representation of $\mathfrak{g}$, but it still has weight decomposition and thus we can define its character). Thus, we need to compute the character of $U\mathfrak{n}_-$. On the other hand, by the PBW theorem (Theorem 5.11), monomials $\prod_{\alpha \in R_+} f_\alpha^{n_\alpha}$ form a basis in $U\mathfrak{n}_-$. Thus,

$$\mathrm{ch}(U\mathfrak{n}_-) = \sum_{\mu \in Q_+} e^{-\mu} P(\mu),$$

where $P(\mu)$ is so-called *Kostant partition function* defined by

$$P(\mu) = \left(\text{number of ways to write } \mu = \sum_{\alpha \in R_+} n_\alpha \alpha \right) \tag{8.22}$$

On the other hand, explicit computation shows that

$$\prod_{\alpha \in R_+} \frac{1}{1 - e^{-\alpha}} = \prod_{\alpha \in R_+} (1 + e^{-\alpha} + e^{-2\alpha} + \dots)$$

$$= \sum_{\mu \in Q_+} P(\mu) e^{-\mu} = \mathrm{ch}(U\mathfrak{n}_-).$$

Now we are ready to give the celebrated Weyl character formula, which gives the characters of irreducible highest weight representations.   □

Theorem 8.34. *Let L_λ be the irreducible finite-dimensional representation with highest weight $\lambda \in P_+$. Then*

$$\mathrm{ch}(L_\lambda) = \frac{\sum_{w \in W} (-1)^{l(w)} e^{w.\lambda}}{\prod_{\alpha \in R_+}(1 - e^{-\alpha})} = \frac{\sum_{w \in W} (-1)^{l(w)} e^{w(\lambda+\rho)}}{\prod_{\alpha \in R_+}(e^{\alpha/2} - e^{-\alpha/2})}$$

where $l(w)$ is the length of $w \in W$ (see Theorem 7.37).

Note that since we already know that L_λ is finite-dimensional the quotient is in fact polynomial, i.e. lies in $\mathbb{C}[P]$ rather than in the completion $\widehat{\mathbb{C}[P]}$.

Proof. We will use the BGG resolution. Recall from linear algebra that if we have a long exact sequence of vector spaces $0 \to V_1 \to \cdots \to V_n \to 0$, then $\sum(-1)^i \dim V_i = 0$. Similarly, if we have a long exact sequence of $\mathfrak{g}$-modules, then applying the previous argument to each weight subspace separately we see that $\sum(-1)^i \mathrm{ch}(V_i) = 0$.

Applying this to the BGG resolution, we see that

$$\mathrm{ch}(L_\lambda) = \sum_{w \in W} (-1)^{l(w)} \, \mathrm{ch}(M_{w.\lambda}).$$

Since characters of Verma modules are given by Lemma 8.33, we get

$$\mathrm{ch}(L_\lambda) = \sum_{w \in W} (-1)^{l(w)} \frac{e^{w.\lambda}}{\prod_{\alpha \in R_+}(1 - e^{-\alpha})}$$

which gives the first form of Weyl character formula. To get the second form, notice that $e^{w.\lambda} = e^{w(\lambda+\rho)-\rho} = e^{-\rho}e^{w(\lambda+\rho)}$ and

$$\prod(1 - e^{-\alpha}) = \prod e^{-\alpha/2}(e^{\alpha/2} - e^{-\alpha/2}) = e^{-\rho}\prod(e^{\alpha/2} - e^{-\alpha/2})$$

since $\rho = \frac{1}{2}\sum \alpha$ (see Lemma 7.36). $\square$

Remark 8.35. There are many proofs of Weyl character formula; in particular, there are several proofs which are "elementary" in that they do not rely on existence of the BGG resolution (see, for example, [22]). However, in our opinion, the BGG resolution, while difficult to prove, provides a better insight into the true meaning of Weyl character formula.

Corollary 8.36 (Weyl denominator identity).

$$\prod_{\alpha \in R_+} (e^{\alpha/2} - e^{-\alpha/2}) = \sum_{w \in W}(-1)^{l(w)}e^{w(\rho)}. \tag{8.23}$$

This polynomial is commonly called Weyl denominator *and denoted by δ.*

Proof. Let us apply Weyl character formula in the case $\lambda = 0$. In this case $L_\lambda = \mathbb{C}$, so $\mathrm{ch}(L_\lambda) = 1$, which immediately gives the statement of the corollary. $\square$

Corollary 8.37. *For $\lambda \in P_+$,*

$$\mathrm{ch}(L_\lambda) = A_{\lambda+\rho}/A_\rho$$

where

$$A_{\mu} = \sum_{w \in W} (-1)^{l(w)} e^{w(\mu)}.$$

Notice that it is immediate from the Weyl denominator identity that the Weyl denominator is skew-symmetric:

$$w(\delta) = (-1)^{l(w)}\delta.$$

Thus, Weyl character formula represents a W-symmetric polynomial $\mathrm{ch}(L_\lambda)$ as a quotient of two skew-symmetric polynomials.

Weyl character formula can also be used to compute the dimensions of irreducible representations. Namely, since

$$\dim V = \sum \dim V[\lambda] = \mathrm{ch}(V)(0)$$

(considering $\mathbb{C}[P]$ as functions on $\mathfrak{h}/2\pi i Q^\vee$ using (8.5)), in theory, dimension of L_λ can be obtained by computing the value of $\mathrm{ch}(L_\lambda)$ at $t = 0$. However, Weyl character formula gives $\mathrm{ch}(L_\lambda)$ as a quotient of two polynomials, both vanishing at $t = 0$; thus, computing the value of the quotient at 0 is not quite trivial. The easiest way to do this is by introducing so-called q-dimension.

Definition 8.38. For a finite-dimensional representation V of $\mathfrak{g}$, define $\dim_q V \in \mathbb{C}[q^{\pm 1}]$ by

$$\dim_q V = \mathrm{tr}_V(q^{2\rho}) = \sum_\lambda (\dim V[\lambda]) q^{2(\rho,\lambda)},$$

where $(\cdot, \cdot)$ is a W-invariant symmetric bilinear form on $\mathfrak{h}^*$ such that $(\lambda, \mu) \in \mathbb{Z}$ for any $\lambda, \mu \in P$.

Obviously, q-dimension can be easily computed from character:

$$\dim_q V = \pi_\rho(\mathrm{ch}(V))$$

where $\pi_\rho: \mathbb{C}[P] \to \mathbb{C}[q^{\pm 1}]$ is the homomorphism defined by $\pi_\rho(e^\lambda) = q^{2(\lambda,\rho)}$. Equally obviously, usual dimension can be easily computed from q-dimension: $\dim V = (\dim_q V)|_{q=1}$.

Theorem 8.39. *For* $\lambda \in P_+$,

$$\dim_q L_\lambda = \prod_{\alpha \in R_+} \frac{q^{(\lambda+\rho,\alpha)} - q^{-(\lambda+\rho,\alpha)}}{q^{(\rho,\alpha)} - q^{-(\rho,\alpha)}},$$

Proof. It follows from the Weyl character formula that

$$\dim_q L_\lambda = \frac{\sum_w (-1)^{l(w)} q^{2(w(\lambda+\rho),\rho)}}{\prod_{\alpha \in R_+} (q^{(\alpha,\rho)} - q^{-(\alpha,\rho)})}$$

The numerator of this expression can be rewritten as follows, using W-invariance of $(\cdot,\cdot)$:

$$\sum_w (-1)^{l(w)} q^{2(w(\lambda+\rho),\rho)} = \sum_w (-1)^{l(w)} q^{2(\lambda+\rho,w(\rho))}$$

$$= \pi_{\lambda+\rho}\left(\sum_w (-1)^{l(w)} e^{w(\rho)}\right)$$

where $\pi_{\lambda+\rho}(e^\mu) = q^{2(\lambda+\rho,\mu)}$.

Using the Weyl denominator identity, we can rewrite this as

$$\pi_\lambda + \rho\left(\prod_{\alpha \in R_+} (e^{\alpha/2}) - e^{(-\alpha/2)}\right) = \prod_{\alpha \in R_+} (q^{(\lambda+\rho,\alpha)} - q^{-(\lambda+\rho,\alpha)})$$

which gives the statement of the theorem. □

Corollary 8.40. *For* $\lambda \in P_+$,

$$\dim L_\lambda = \prod_{\alpha \in R_+} \frac{(\lambda + \rho, \alpha)}{(\rho, \alpha)} = \prod_{\alpha \in R_+} \frac{\langle \lambda + \rho, \alpha^\vee \rangle}{\langle \rho, \alpha^\vee \rangle}.$$

Proof. Follows from

$$\lim_{q \to 1} \frac{q^n - q^{-n}}{q^m - q^{-m}} = \frac{n}{m},$$

which can be verified either by noticing that

$$\frac{q^n - q^{-n}}{q - q^{-1}} = q^{n-1} + q^{n-3} + \cdots + q^{-n+1}$$

or by using L'Hôpital's rule. □

It should be noted that explicitly computing characters using Weyl character formula can lead to extremely long computations (suffices to mention that the Weyl group of type E_8 has order $696,729,600$). There are equivalent formulas which are slightly more convenient for computations, such a Freudental's formula (see [22]); however, with any of these formulas doing computations by hand is extremely tedious. Fortunately, there are software packages which allow one to delegate this job to a computer. Among the most popular are the `weyl` package for Maple, developed by John Stembridge [56], the `LiE` program developed by Marc van Leeuwen [35], and the `GAP` computational discrete algebra system [13].

8.6. Multiplicities

Since finite-dimensional irreducible representations of $\mathfrak{g}$ are classified by dominant weights $\lambda \in P_+$, it follows from complete reducibility that any finite-dimensional representation can be written as

$$V = \bigoplus_{\lambda \in P_+} n_\lambda L_\lambda. \tag{8.24}$$

In this section, we discuss how one can compute multiplicities n_λ.

Theorem 8.41. *Characters* $\mathrm{ch}(L_\lambda)$, $\lambda \in P_+$, *form a basis in the algebra of W-invariant polynomials* $\mathbb{C}[P]^W$.

Proof. First, note that we have a fairly obvious basis in $\mathbb{C}[P]^W$. Namely, for any $\lambda \in P_+$ let

$$m_\lambda = \sum_{\mu \in W\lambda} \mathrm{e}^\mu,$$

where $W\lambda$ is the W-orbit of λ. Since any orbit contains a unique element of P_+ (Lemma 8.22), it is clear that elements m_λ, $\lambda \in P_+$, form a basis in $\mathbb{C}[P]^W$.

It follows from Theorem 8.15 that for any $\lambda \in P_+$, we have

$$\mathrm{ch}(L_\lambda) = \sum_{\mu \preceq \lambda} c_\mu \mathrm{e}^\mu = m_\lambda + \sum_{\mu \in P_+, \mu \prec \lambda} c_\mu m_\mu,$$

where the coefficients c_μ can be computed using Weyl character formula. Note that for any $\lambda \in P_+$, the set $\{\mu \in P_+ \mid \mu \preceq \lambda\}$ is finite (Exercise 8.2).

Thus, the matrix expressing $\mathrm{ch}(L_\lambda)$ in terms of m_μ is upper-triangular (with respect to partial order $\prec$) with ones on the diagonal. From this, standard linear algebra arguments show that this matrix is invertible:

$$m_\lambda = \operatorname{ch}(L_\lambda) + \sum_{\mu \in P_+, \mu \prec \lambda} d_\mu \operatorname{ch}(L_\mu).$$

$\square$

This theorem shows that multiplicities n_λ in (8.24) can be found by writing character $\operatorname{ch}(V)$ in the basis $\operatorname{ch}(L_\lambda)$:

$$\operatorname{ch}(V) = \sum_{\lambda \in P_+} n_\lambda \operatorname{ch}(L_\lambda).$$

Moreover, the proof of the theorem also suggests a way of finding these coefficients recursively: if $\lambda \in P(V)$ is maximal (i.e., there are no weights $\mu \in P(V)$ with $\lambda \prec \mu$), then $n_\lambda = \dim V[\lambda]$. Now we can consider $\operatorname{ch}(V) - n_\lambda \operatorname{ch}(L_\lambda)$ and apply the same construction, and so on.

For the simplest Lie algebras such as $\mathfrak{sl}(2, \mathbb{C})$, it is easy to find the coefficients explicitly (see Exercise 4.11). For higher-dimensional Lie algebras, computations can be very long and tedious. As with the Weyl character formula, use of a computer package is recommended in such cases.

8.7. Representations of $\mathfrak{sl}(n, \mathbb{C})$

In this section, we will consider in detail the classification of irreducible representations of $\mathfrak{sl}(n, \mathbb{C})$ and the character formula for irreducible representations of $\mathfrak{sl}(n, \mathbb{C})$.

We start by recalling the root system of $\mathfrak{sl}(n, \mathbb{C})$ (see Example 6.40, Example 7.4). In this case the root system is given by

$$R = \{e_i - e_j, i \neq j\} \subset \mathfrak{h}^* = \mathbb{C}^n / \mathbb{C}(1, \ldots, 1)$$

and positive roots are $e_i - e_j, i < j$. The weight lattice and set of dominant roots are given by

$$P = \{(\lambda_1, \ldots, \lambda_n) \in \mathfrak{h}^* \mid \lambda_i - \lambda_j \in \mathbb{Z}\}$$
$$P_+ = \{(\lambda_1, \ldots, \lambda_n) \in \mathfrak{h}^* \mid \lambda_i - \lambda_{i+1} \in \mathbb{Z}_+\}.$$

Note that since adding a multiple of $(1, \ldots, 1)$ does not change the weight, we can represent every weight $\lambda \in P$ by an n-tuple $(\lambda_1, \ldots, \lambda_n)$ such that each

$\lambda_i \in \mathbb{Z}$. Similarly,

$$P_+ = \{(\lambda_1, \ldots, \lambda_n) \mid \lambda_i \in \mathbb{Z}, \lambda_1 \geq \lambda_2 \geq \cdots \geq \lambda_n\}/\mathbb{Z}(1, \ldots, 1)$$
$$= \{(\lambda_1, \ldots, \lambda_{n-1}, 0) \mid \lambda_i \in \mathbb{Z}_+, \lambda_1 \geq \lambda_2 \geq \cdots \geq \lambda_{n-1} \geq 0\}.$$

(For readers familiar with the notion of partition, we note that the last formula shows that the set of dominant integer weights for $\mathfrak{sl}(n, \mathbb{C})$ can be identified with the set of partitions with $n - 1$ parts.)

It is common to represent dominant weights graphically by so-called *Young diagrams*, as illustrated here.

$$(5, 3, 1, 1, 0) \longrightarrow$$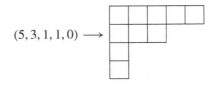

More generally, a Young diagram corresponding to a weight $(\lambda_1 \geq \cdots \geq \lambda_{n-1} \geq 0)$ is constructed by putting λ_1 boxes in the first row, λ_2 boxes in the second row, and so on.

Example 8.42. Let $V = \mathbb{C}^n$ be the tautological representation of $\mathfrak{sl}(n, \mathbb{C})$. Then weights of V are $e_1, \ldots, e_n$. One easily sees that the highest weight is $e_1 = (1, 0, \ldots, 0)$, so $V = L_{(1,0,\ldots,0)}$. The corresponding Young diagram is a single box.

Example 8.43. Let $k \geq 0$. Then it can be shown (see Exercise 8.4) that the representation $S^k \mathbb{C}^n$ is a highest weight representation with highest weight $ke_1 = (k, 0, \ldots, 0)$. The corresponding Young diagram is a row of k boxes.

Example 8.44. Let $1 \leq k < n$. Then it can be shown (see Exercise 8.5) that the representation $\Lambda^k \mathbb{C}^n$ is a highest weight representation with highest weight $e_1 + \cdots + e_k = (1, 1, \ldots, 1, 0, \ldots 0)$. The corresponding Young diagram is a column of k boxes.

Note that the same argument shows that for $k = n$, the highest weight of $\Lambda^n \mathbb{C}^n$ is $(1, \ldots, 1) = (0, \ldots, 0)$, so $\Lambda^n \mathbb{C}^n$ is the trivial one-dimensional representation of $\mathfrak{sl}(n, \mathbb{C})$ (compare with Exercise 4.3).

In fact, the above examples can be generalized: any irreducible finite-dimensional representation L_λ of $\mathfrak{sl}(n, \mathbb{C})$ appears as a subspace in $(\mathbb{C}^n)^N$, $N = \sum \lambda_i$, determined by suitable symmetry properties, i.e. transforming in a certain way under the action of symmetric group S_N. Detailed exposition can be found in [11].

Example 8.45. Let V be the adjoint representation of $\mathfrak{sl}(3, \mathbb{C})$. Then the highest weight of V is $\alpha_1 + \alpha_2 = e_1 - e_3 = 2e_1 + e_2$. Thus, $V = L_{(2,1,0)}$ and the corresponding Young diagram is

We can also give an explicit description of the algebra $\mathbb{C}[P]$. Namely, denoting $x_i = e^{e_i}$, we get $e^\lambda = x_1^{\lambda_1} \ldots x_n^{\lambda_n}$. Relation $e_1 + \cdots + e_n = 0$ gives $x_1 x_2 \ldots x_n = 1$. Thus,

$$\mathbb{C}[P] = \mathbb{C}[x_1^{\pm 1}, \ldots, x_n^{\pm 1}]/(x_1 \ldots x_n - 1). \tag{8.25}$$

It is easy to check that two *homogeneous* polynomials of the same total degree are equal in $\mathbb{C}[P]$ iff they are equal in $\mathbb{C}[x_1^{\pm 1}, \ldots, x_n^{\pm 1}]$.

Let us now discuss the characters of irreducible representations of $\mathfrak{sl}(n, \mathbb{C})$. We start by writing the Weyl denominator identity in this case.

Theorem 8.46. *The Weyl denominator identity for* $\mathfrak{sl}(n, \mathbb{C})$ *takes the form*

$$\prod_{i<j}(x_i - x_j) = \sum_{s \in S_n} \operatorname{sgn}(s) x_{s(1)}^{n-1} x_{s(2)}^{n-1} \ldots x_{s(n)}^0, \tag{8.26}$$

where $\operatorname{sgn}(s) = (-1)^{l(s)}$ *is the sign of permutation s.*

Proof. Using $\rho = (n-1, n-2, \ldots, 1, 0)$, we can write the left-hand side of the Weyl denominator identity (8.23) as

$$e^\rho \prod_{\alpha \in R_+}(1 - e^{-\alpha}) = x_1^{n-1} x_2^{n-2} \ldots x_n^0 \prod_{i<j}\left(1 - \frac{x_j}{x_i}\right) = \prod_{i<j}(x_i - x_j).$$

The right-hand side is

$$\sum_{s \in S_n}(-1)^{l(s)} s(x_1^{n-1} \ldots x_n^0) = \sum_{s \in S_n} \operatorname{sgn}(s) x_{s(1)}^{n-1} \ldots x_{s(n)}^0.$$

$\square$

Of course, identity (8.26) is well-known: most readers probably have recognized that the right-hand side is the well-known Vandermonde determinant, i.e. the determinant of the matrix $(x_j^{n-i})_{1 \le i, j \le n}$. Formula (8.26) for Vandermonde determinant is usually discussed in any standard algebra course and can be proved by completely elementary methods.

Now we are ready to discuss the Weyl character formula.

Theorem 8.47. *Let* $\lambda = (\lambda_1, \ldots, \lambda_n,) \in P_+$ *be a dominant weight for* $\mathfrak{sl}(n, \mathbb{C})$*:* $\lambda_i \in \mathbb{Z}_+, \lambda_1 \geq \lambda_2 \geq \cdots \geq \lambda_n$ *(we do not assume that* $\lambda_n = 0$*). Then the character of the corresponding irreducible representation of* $\mathfrak{sl}(n, \mathbb{C})$ *is given by*

$$\text{ch}(L_\lambda) = \frac{A_{\lambda_1+n-1,\lambda_2+n-2,\ldots,\lambda_n}}{A_{n-1,n-2,\ldots,1,0}} = \frac{A_{\lambda_1+n-1,\lambda_2+n-2,\ldots,\lambda_n}}{\prod_{i<j}(x_i - x_j)}, \qquad (8.27)$$

where

$$A_{\mu_1,\ldots,\mu_n} = \det(x_i^{\mu_j})_{1\leq i,j\leq n} = \sum_{s\in S_n} \text{sgn}(s)x_{s(1)}^{\mu_1} \cdots x_{s(n)}^{\mu_n}.$$

Proof. This immediately follows from the general Weyl character formula (Theorem 8.34), together with $\rho = (n - 1, \ldots, 1, 0)$. $\qquad \square$

Polynomials (8.27) are usually called *Schur functions* and denoted s_λ. It follows from the general result about W-invariance of characters (Theorem 8.8) that s_λ are symmetric polynomials in $x_1, \ldots, x_n$; moreover, by Theorem 8.41, they form a basis of the space of symmetric polynomials. A detailed description of these functions can be found, for example, in Macdonald's monograph [36].

Example 8.48. Let us write the Weyl character formula for $\mathfrak{sl}(3, \mathbb{C})$. In this case, $W = S_3$, so the Weyl character formula gives for $\lambda = (\lambda_1, \lambda_2, 0), \lambda_1 \geq \lambda_2$:

$$\text{ch}(L_\lambda) = \frac{\sum_{s\in S_3} \text{sgn}(s)x_{s(1)}^{\lambda_1+2} x_{s(2)}^{\lambda_2+1}}{\prod_{i<j}(x_i - x_j)}.$$

Let us check this formula for the fundamental representation, i.e. the tautological action of $\mathfrak{sl}(3, \mathbb{C})$ on $\mathbb{C}^3$. In this case, weights of this representations are $e_1 = (1, 0, 0)$, $e_2 = (0, 1, 0)$, $e_3 = (0, 0, 1)$, so the highest weight is $(1, 0, 0)$. Therefore, Weyl character formula gives

$$\text{ch}(\mathbb{C}^3) = \frac{\sum_{s\in S_3} \text{sgn}(s)s(x_1^3 x_2)}{(x_1 - x_2)(x_1 - x_3)(x_2 - x_3)}$$

$$= \frac{x_1^3 x_2 - x_1^3 x_3 - x_2^3 x_1 + x_2^3 x_3 + x_3^3 x_1 - x_3^3 x_2}{(x_1 - x_2)(x_1 - x_3)(x_2 - x_3)}.$$

We leave it to the reader to check that the rational function in the right-hand side is actually equal to $x_1 + x_2 + x_3$, thus showing that the weights of $\mathbb{C}^3$ are e_1, e_2, e_3, each with multiplicity one.

8.8. Harish–Chandra isomorphism

Recall that in Section 6.3 we have defined a central element $C \in Z(U\mathfrak{g})$, called the Casimir element. This element played an important role in the proof of the complete reducibility theorem.

However, the Casimir element is not the only central element in $U\mathfrak{g}$. In this section, we will study the center

$$Z\mathfrak{g} = Z(U\mathfrak{g}).$$

In particular, we will show that central elements can be used to distinguish finite-dimensional representations: an irreducible finite-dimensional representation V is completely determined by the values of the central elements $C \in Z\mathfrak{g}$ in V.

We start by recalling some results about $U\mathfrak{g}$ that were proved in Section 5.2. Recall that for any vector space V we denote by SV the symmetric algebra of V; it can be identified with the algebra of polynomial functions on V^*. In particular, we denote by $S\mathfrak{g}$ the symmetric algebra of $\mathfrak{g}$. By Theorem 5.16, the map

$$\mathbf{sym} \colon S\mathfrak{g} \to U\mathfrak{g}$$

$$x_1 \ldots x_n \mapsto \frac{1}{n!} \sum_{s \in S_n} x_{s(1)} \ldots x_{s(n)} \tag{8.28}$$

is an isomorphism of $\mathfrak{g}$-modules, compatible with natural filtrations in $S\mathfrak{g}$, $U\mathfrak{g}$. Note, however, that $\mathbf{sym}$ is not an algebra isomorphism – it cannot be, because $S\mathfrak{g}$ is commutative and $U\mathfrak{g}$ is not (unless $\mathfrak{g}$ is abelian).

Proposition 8.49. *Map* $\mathbf{sym}$ *induces a vector space isomorphism*

$$(S\mathfrak{g})^G \xrightarrow{\sim} Z\mathfrak{g},$$

where G is the connected simply-connected Lie group with Lie algebra $\mathfrak{g}$ and $Z\mathfrak{g}$ is the center of $U\mathfrak{g}$.

Proof. Indeed, it was proved in Proposition 5.7 that $Z\mathfrak{g}$ coincides with the subspace of $\mathfrak{g}$–invariants in $U\mathfrak{g}$. On the other hand, for any representation of a connected Lie group G, spaces of G-invariants and $\mathfrak{g}$-invariants coincide. $\quad\square$

Note that the map $\mathbf{sym} \colon (S\mathfrak{g})^G \xrightarrow{\sim} Z\mathfrak{g}$ is not an algebra isomorphism (even though both algebras are commutative); we will return to this below.

Thus, central elements in $U\mathfrak{g}$ are closely related to invariant polynomials $(S\mathfrak{g})^G$, which makes them much easier to describe. However, it turns out that the same space admits an even more explicit description.

Choose a Cartan subalgebra $\mathfrak{h} \subset \mathfrak{g}$ and consider the algebra $S\mathfrak{h}$ of polynomials on $\mathfrak{h}^*$. Since $\mathfrak{h}$ is a direct summand in $\mathfrak{g}$: $\mathfrak{g} = \mathfrak{h} \oplus \bigoplus_\alpha \mathfrak{g}_\alpha$, we see that $\mathfrak{h}^*$ is a direct summand in $\mathfrak{g}^*$. Thus, we can restrict any polynomial $p \in S\mathfrak{g}$ to $\mathfrak{h}^*$. This gives a restriction map

$$\mathbf{res}\colon S\mathfrak{g} \to S\mathfrak{h}. \tag{8.29}$$

It is easy to see that **res** is a degree-preserving algebra homomorphism.

In particular, we can apply **res** to a G–invariant polynomial $p \in (S\mathfrak{g})^G$. Since the coadjont action of G does not preserve $\mathfrak{h}^* \subset \mathfrak{g}^*$, we can not claim that the restriction of p to $\mathfrak{h}^*$ is G-invariant. However, there are some inner automorphisms which preserve $\mathfrak{h}^*$: for example, we have seen in Exercise 7.16 that any element of the Weyl group can be lifted to an inner automorphism of $\mathfrak{g}^*$, i.e. is given by $\mathrm{Ad}^* \tilde{w}$ for some $\tilde{w} \in G$. Thus, we see that restriction map (8.29) gives rise to a map **res**$\colon (S\mathfrak{g})^G \to (S\mathfrak{h})^W$.

Theorem 8.50. *Restriction map* (8.29) *induces an algebra isomorphism*

$$\mathbf{res}\colon (S\mathfrak{g})^G \to (S\mathfrak{h})^W. \tag{8.30}$$

The proof of this theorem can be found in [22] or [9]. Here we only note that to prove surjectivity, we need to construct sufficiently many G-invariant elements in $S\mathfrak{g}$, which is done using irreducible finite-dimensional representations.

Combining the results above, we see that we have the following diagram

$$\tag{8.31}$$

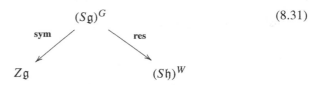

where both arrows are isomorphisms: **sym** is an isomorphism of filtered vector spaces, while **res** is an isomorphism of graded algebras.

Example 8.51. Let $\Omega = \sum a_i \otimes b_i \in (S^2\mathfrak{g})^G$ be an invariant symmetric tensor. Then

$$\mathbf{sym}(\Omega) = \frac{1}{2}\left(\sum a_i b_i + b_i a_i\right) = \sum a_i b_i.$$

In particular, if B is a non-degenerate invariant symmetric bilinear form on $\mathfrak{g}$ and x_i, x^i are dual bases with respect to B, then we can take $\Omega = \sum x_i \otimes x^i$ and

sym$(\Omega) = \sum x_i x^i \in U\mathfrak{g}$ is exactly the corresponding Casimir element C_B as defined in Proposition 6.15.

Returning to the diagram (8.31), we see that the composition **res** $\circ$ (**sym**)$^{-1}$ gives an isomorphism $Z\mathfrak{g} \xrightarrow{\sim} (S\mathfrak{h})^W$, which makes it easy to describe how large $Z\mathfrak{g}$ is as a filtered vector space. However, it is not an algebra isomorphism. A natural question is whether one can identify $Z\mathfrak{g}$ and $(S\mathfrak{h})^W$ as algebras.

To answer that, we will consider action of central elements $z \in Z\mathfrak{g}$ in highest weight representations.

Theorem 8.52. *For any $z \in Z\mathfrak{g}$, there exists a unique polynomial $\chi_z \in S\mathfrak{h}$ such that in any highest weight representation V with highest weight λ,*

$$z|_V = \chi_z(\lambda + \rho)\,\mathrm{id}\,. \tag{8.32}$$

The map $z \mapsto \chi_z$ is an algebra homomorphism $Z\mathfrak{g} \to (S\mathfrak{h})^W$.

Proof. Since any $z \in Z\mathfrak{g}$ must have weight zero (which follows because it must be ad $\mathfrak{h}$ invariant), we see that if v_λ is the highest weight vector of a highest weight representation V, then $zv_\lambda = cv_\lambda$ for some constant c. Since z is central and v_λ generates V, this implies that $z = c$ id in V for some constant c which depends on λ and which therefore can be written as $\chi_z(\lambda + \rho)$ for some function χ_z on $\mathfrak{h}^*$.

To show that χ_z is a polynomial in λ, we extend the definition of χ_z to all of $U\mathfrak{g}$ as follows. Recall that by the PBW theorem, monomials

$$\Big(\prod_{\alpha \in R_+} f_\alpha^{k_\alpha} \Big) \Big(\prod_i h_i^{n_i} \Big) \Big(\prod_{\alpha \in R_+} e_\alpha^{m_\alpha} \Big),$$

where h_i is a basis in $\mathfrak{h}$, form a basis in $U\mathfrak{g}$. Define now the map $HC \colon U\mathfrak{g} \to S\mathfrak{h}$ by

$$
HC\Big(\Big(\prod_{\alpha \in R_+} f_\alpha^{k_\alpha} \Big) \Big(\prod_i h_i^{n_i} \Big) \Big(\prod_{\alpha \in R_+} e_\alpha^{m_\alpha} \Big) \Big)
$$
$$
= \begin{cases} \prod_i h_i^{n_i}, & \text{if } k_\alpha, m_\alpha = 0 \text{ for all } \alpha \\ 0 & \text{otherwise} \end{cases} \tag{8.33}
$$

Then an easy explicit computation shows that for any $u \in U\mathfrak{g}$, $uv_\lambda = HC(u)(\lambda)v_\lambda + \cdots$, where dots stand for vectors of weights $\mu \prec \lambda$. In particular, applying this to $z \in Z\mathfrak{g}$, we see that $zv_\lambda = \chi_z(\lambda + \rho)v_\lambda = HC(z)(\lambda)v_\lambda$, so $\chi_z(\lambda) = HC(z)(\lambda - \rho)$. By definition, $HC(z) \in S\mathfrak{h}$, so $\chi_z(\lambda)$ is a polynomial function in λ.

The fact that $z \mapsto \chi_z$ is an algebra homomorphism is obvious:

$$(z_1 z_2)v_\lambda = z_1(z_2 v_\lambda) = \chi_{z_2}(\lambda + \rho)z_1 v_\lambda = \chi_{z_1}(\lambda + \rho)\chi_{z_2}(\lambda + \rho)v_\lambda$$

so $\chi_{z_1 z_2} = \chi_{z_1}\chi_{z_2}$, and similarly for addition.

Finally, we need to show that for every $z \in Z\mathfrak{g}$, χ_z is W-invariant. It suffices to show that $\chi_z(s_i(\lambda)) = \chi_z(\lambda)$ for any i.

Let $\lambda \in P_+$. Then, as was shown in Theorem 8.23, the Verma module M_λ contains a submodule $M_{s_i.\lambda}$, where $s_i.\lambda = s_i(\lambda + \rho) - \rho$. Therefore, the value of z in M_λ and $M_{s_i.\lambda}$ must be equal, which gives

$$\chi_z(\lambda + \rho) = \chi_z(s_i.\lambda + \rho) = \chi_z(s_i(\lambda + \rho)).$$

Thus, we see that $\chi_z(\mu) = \chi_z(s_i(\mu))$ for any $\mu \in \rho + P_+$. However, since both $\chi_z(\mu)$ and $\chi_z(s_i(\mu))$ are polynomial functions of μ, it is easy to show that if they are equal for all $\mu \in \rho + P_+$, then they are everywhere equal (see Exercise 8.8). Thus, χ_z is s_i-invariant. $\qquad \square$

Example 8.53. Let $(\,,\,)$ be an invariant symmetric bilinear form on $\mathfrak{g}^*$ and C the corresponding Casimir element as in Example 8.51. Then explicit computation, done in Exercise 8.7, shows that the action of C in a highest weight module with highest weight λ is given by $(\lambda, \lambda + 2\rho) = (\lambda + \rho, \lambda + \rho) - (\rho, \rho)$. Therefore, $\chi_C(\mu) = (\mu, \mu) - (\rho, \rho)$.

We can now add the map $z \to \chi$ to the diagram (8.31):

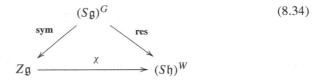

$$(8.34)$$

Note, however, that the diagram is not commutative. For example, for an invariant bilinear form $(\,,\,)$ on $\mathfrak{g}^*$, considered as an element of $(S^2\mathfrak{g})^G$, composition $\chi \circ \mathbf{sym}$ gives the polynomial $(\mu, \mu) - (\rho, \rho)$ (see Example 8.53), whereas the restriction gives just (μ, μ).

Theorem 8.54.

(1) *Diagram* (8.34) *is commutative "up to lower order terms": for any* $p \in (S^n\mathfrak{g})^G$, *we have*

$$\chi_{\mathbf{sym}(p)} \equiv \mathbf{res}(p) \quad \mathrm{mod}\ (S^{n-1}\mathfrak{h})^W$$

(2) *The map* $\chi : Z\mathfrak{g} \rightarrow (S\mathfrak{h})^W$ *defined in Theorem 8.52 is an alge-bra isomorphism. This isomorphism is usually called* Harish–Chandra *isomorphism.*

Proof. We start with part (1). It is easier to prove a more general result: for any $p \in S^n\mathfrak{g}$, we have

$$HC(\mathbf{sym}(p)) \equiv \mathbf{res}(p) \mod S^{n-1}\mathfrak{h},$$

where $HC : U\mathfrak{g} \rightarrow S\mathfrak{h}$ is defined by (8.33).

Indeed, since $\mathbf{sym}(x_1 \dots x_n) \equiv x_1 \dots x_n \mod U_{n-1}\mathfrak{g}$, we see that if

$$p = \Big(\prod_{\alpha \in R_+} f_\alpha^{k_\alpha} \Big)\Big(\prod_i h_i^{n_i} \Big)\Big(\prod_{\alpha \in R_+} e_\alpha^{m_\alpha} \Big) \in S^n\mathfrak{g}$$

then we have

$$HC(\mathbf{sym}(p)) \equiv \begin{cases} \prod_i h_i^{n_i}, & k_\alpha, m_\alpha = 0 \text{ for all } \alpha \\ 0 & \text{otherwise} \end{cases} = \mathbf{res}(p) \mod S^{n-1}\mathfrak{h}.$$

Since for $z \in Z\mathfrak{g} \cap U_n\mathfrak{g}$ we have $\chi_z(\lambda) = HC(z)(\lambda - \rho) \equiv HC(z)(\lambda)$ mod $S^{n-1}\mathfrak{h}$ (see proof of Theorem 8.52), we see that $\chi_z \equiv HC(z) \mod S^{n-1}\mathfrak{h}$, which proves part (1).

To prove part (2), note that since $\mathbf{res}$ is an isomorphism, part (1) implies that composition $\chi \circ \mathbf{sym}$ is also an isomorphism. Since $\mathbf{sym}$ is an isomorphism, χ is also an isomorphism. $\qquad\square$

Corollary 8.55. *Let* $\lambda, \mu \in \mathfrak{h}^*$. *Then* $\chi_z(\lambda) = \chi_z(\mu)$ *for all* $z \in Z\mathfrak{g}$ *iff* λ, μ *are in the same* W-*orbit.*

Indeed it follows from the previous theorem and the fact that W-invariant polynomials separate orbits of W.

Theorem 8.54 also allows one to construct an algebra isomorphism $(S\mathfrak{g})^G \xrightarrow{\sim} Z\mathfrak{g}$ as a composition $\chi^{-1} \circ \mathbf{res}$. In fact, it is a special case of a more general result: for any finite-dimensional Lie algebra $\mathfrak{g}$ there exists an algebra isomorphism $(S\mathfrak{g})^G \xrightarrow{\sim} Z\mathfrak{g}$, called *Duflo–Kirillov* (or, more properly, Duflo–Ginzburg–Kirillov) isomorphism. For semisimple Lie algebras this isomorphism coincides with the one given by composition $\chi^{-1} \circ \mathbf{res}$. Details can be found in [9].

Returning to the case of semisimple $\mathfrak{g}$, it should be noted that the algebra $(S\mathfrak{h})^W$ (and thus $Z\mathfrak{g}$) is well studied (see, e.g., [23]). Namely, it is known that this algebra is a free polynomial algebra, with the number of generators equal to the rank of $\mathfrak{g}$:

$$(S\mathfrak{h})^W \simeq \mathbb{C}[C_1, \ldots, C_r], \quad r = \mathrm{rank}(\mathfrak{g}).$$

Degrees of the generators are also known. For various reasons it is common to consider not degrees themselves but so-called *exponents* of $\mathfrak{g}$ (or of W)

$$d_i = \deg C_i - 1.$$

For example, for $\mathfrak{g} = \mathfrak{sl}(n, \mathbb{C})$ we have $(S\mathfrak{h})^W = (\mathbb{C}[x_1, \ldots, x_n]/(x_1 + \cdots + x_n))^{S_n} = \mathbb{C}[\sigma_2, \ldots, \sigma_n]$, where σ_i are elementary symmetric functions, i.e. coefficients of the polynomial $\prod(x - x_i)$. Thus, in this case the exponents are $1, \ldots, n - 1$. Lists of exponents for other simple Lie algebras can be found in [3]. We only mention here that existence of Killing form implies that $d_1 = 1$ for any simple Lie algebra.

Exponents also appear in many other problems related to semisimple Lie algebras. For example, it is known that if G is a compact real semisimple Lie group, then (topological) cohomology of G is a free exterior algebra:

$$H^*(G, \mathbb{R}) \simeq \Lambda[\omega_1, \ldots, \omega_r], \quad \deg \omega_i = 2d_i + 1,$$

where d_i are the exponents of $\mathfrak{g}_{\mathbb{C}}$. For example, generator $\omega_1 \in H^3(G)$ which corresponds to $d_1 = 1$ is defined (up to a scalar) by Exercise 4.7. Detailed discussion of this and related topics can be found in [14].

8.9. Proof of Theorem 8.25

In this section, we give a proof of Theorem 8.25. Recall the statement of the theorem.

Theorem. *Let* $\lambda \in P_+$, *and let*

$$\tilde{L}_\lambda = M_\lambda / \sum M_i,$$

where $M_i \subset M_\lambda$ *is the subrepresentation generated by vector* $v_{s_i.\lambda} = f_i^{n_i+1} v_\lambda$, $n_i = \langle \lambda, \alpha_i^\vee \rangle$. *Then* $\tilde{L}_\lambda$ *is finite-dimensional.*

The proof is based on the following important notion. Recall that $\mathfrak{sl}(2, \mathbb{C})_i$ is the subalgebra generated by e_i, f_i, h_i, see Lemma 6.42.

Definition 8.56. A representation of $\mathfrak{g}$ is called *integrable* if for any $v \in V$ and any $i \in \{1, \ldots, r\}$, the $\mathfrak{sl}(2, \mathbb{C})_i$-submodule generated by v is finite-dimensional:

$$\dim(U\mathfrak{sl}(2, \mathbb{C})_i v) < \infty.$$

The theorem itself follows from the following two lemmas.

Lemma 8.57. *For any representation V, let*

$$V^{\text{int}} = \{v \in V \mid \textit{For any } i, \ \dim(U\mathfrak{sl}(2, \mathbb{C})_i v) < \infty\} \subset V.$$

Then V^{int} is an integrable subrepresentation of V.

Lemma 8.58. *Any highest weight integrable representation is finite-dimensional.*

From these two lemmas, the theorem easily follows. Indeed, consider $\tilde{L}_\lambda = M_\lambda / \sum M_i$. Since in $\tilde{L}_\lambda$, $f_i^{n_i+1} v_\lambda = 0$ and $e_i v_\lambda = 0$, it is easy to see that v_λ generates a finite-dimensional $U\mathfrak{sl}(2, \mathbb{C})_i$–module and thus $v_\lambda \in \tilde{L}_\lambda^{\text{int}}$. Since v_λ generates $\tilde{L}_\lambda$, it follows from Lemma 8.57 that $\tilde{L}_\lambda$ is integrable. By Lemma 8.58, this implies that $\tilde{L}_\lambda$ is finite-dimensional.

Thus, it remains to prove these two lemmas.

Proof of Lemma 8.57. Let $v \in V^{\text{int}}$ and let W be the $\mathfrak{sl}(2, \mathbb{C})_i$–module generated by v. By assumption, W is finite-dimensional. Consider now the vector space $\mathfrak{g}W$, spanned by vectors xw, $x \in \mathfrak{g}$, $w \in W$. Clearly, $\mathfrak{g}W$ is finite-dimensional. It is also closed under the action of $\mathfrak{sl}(2, \mathbb{C})_i$:

$$e_i xw = xe_i w + [x, e_i]w \in \mathfrak{g}W$$

and similarly for other elements of $\mathfrak{sl}(2, \mathbb{C})_i$. Thus, we see that for any $x \in \mathfrak{g}$, xv is contained in the finite-dimensional $\mathfrak{sl}(2, \mathbb{C})_i$–module $\mathfrak{g}W$. Repeating this for all i, we see that $xv \in V^{\text{int}}$. Therefore, V^{int} is a subrepresentation; by definition, it is integrable. $\square$

Proof of Lemma 8.58. Let V be an integrable representation. Since any vector is contained in a finite-dimensional $\mathfrak{sl}(2, \mathbb{C})_i$–submodule, the same arguments as in the proof of Theorem 8.8 show that the set of weights of V is W-invariant. If we additionally assume that V is a highest weight representation with highest weight λ, then any weight μ of V must satisfy $w(\mu) \in \lambda - Q_+$ for any $w \in W$. In particular, choosing w so that $w(\mu) \in P_+$ (which is always possible by Lemma 8.22), we see that $w(\mu) \in P_+ \cap (\lambda - Q_+)$. But for any λ, the set $P_+ \cap (\lambda - Q_+)$ is finite (see Exercise 8.2). Thus, the set of weights of V is finite. Since in a highest weight module, any weight subspace is finite-dimensional, this shows that V is finite-dimensional. $\square$

8.10. Exercises

8.1. Prove Lemma 8.16. You can do it by breaking it into several steps as shown below.

  (1) Show that given any finite set of distinct weights $\lambda_1, \ldots, \lambda_n \in P(V)$, there exists an element $p \in U\mathfrak{h}$ such that $p(\lambda_1) = 1, p(\lambda_i) = 0$ for $i \neq 1$ (considering elements of $U\mathfrak{h} = S\mathfrak{h}$ as polynomial functions on $\mathfrak{h}^*$).

  (2) Let $V \subset M$ be an $\mathfrak{h}$-submodule, and $v \in V$. Write $v = \sum v_i, v_i \in M[\lambda_i]$. Show that then each of $v_i \in V$.

  (3) Deduce Lemma 8.16.

8.2. (1) Show that for any $t \in \mathbb{R}_+$, the set $\{\lambda \in Q_+ \mid (\lambda, \rho) \leq t\}$ is finite.

  (2) Show that for any $\lambda \in P_+$, the set $\{\mu \in P_+ \mid \mu \preceq \lambda\}$ is finite.

8.3. Let $\omega_i, i = 1, \ldots, r = \operatorname{rank} \mathfrak{g}$, be a basis of P, and denote $x_i = e^{\omega_i} \in \mathbb{C}[P]$. Show that then $\mathbb{C}[P]$ is isomorphic to the algebra of Laurent polynomials $\mathbb{C}[x_1^{\pm 1}, \ldots, x_r^{\pm 1}]$.

8.4. Let $k > 0$. Consider the representation $V = S^k \mathbb{C}^n$ of $\mathfrak{sl}(n, \mathbb{C})$.

  (1) Compute all weights of V and describe the corresponding weight subspaces.

  (2) Show that V contains a unique (up to a factor) vector v such that $\mathfrak{n}_+ v = 0$, namely $v = x_1^k$, and deduce from this that V is irreducible.

  (3) Find the highest weight of V and draw the corresponding Young diagram.

8.5. Let $1 \leq k \leq n$. Consider the representation $V = \Lambda^k \mathbb{C}^n$ of $\mathfrak{sl}(n, \mathbb{C})$.

  (1) Compute all weights of V and describe the corresponding weight subspaces.

  (2) Show that V contains a unique (up to a factor) vector v such that $\mathfrak{n}_+ v = 0$, namely $v = x_1 \wedge \cdots \wedge x_k$, and deduce from this that V is irreducible.

  (3) Find the highest weight of V and draw the corresponding Young diagram.

8.6. Let $\mathfrak{g}$ be a simple Lie algebra and let $\theta \in R_+$ be the maximal root of $\mathfrak{g}$ as defined in Example 8.26.

  (1) Show that any $\alpha \in R \cup \{0\}$ can be written in the form $\alpha = \theta - \sum n_i \alpha_i$, $n_i \in \mathbb{Z}_+$, and that this condition uniquely defines θ.

  (2) Show that $\operatorname{ht}(\theta)$ is maximal possible: for any $\alpha \in R_+, \alpha \neq \theta$, we have $\operatorname{ht}(\alpha) < \operatorname{ht}(\theta)$ (here ht is the height of a root, see (7.8)). The number $h = \operatorname{ht}(\theta) + 1$ is called the *Coxeter number* of $\mathfrak{g}$.

8.7. Let $\mathfrak{g}$ be a simple complex Lie algebra and $(\,,\,)$ a non-degenerate invariant bilinear symmetric form on $\mathfrak{g}$. We will also use the same notation $(\,,\,)$ for the corresponding bilinear form on $\mathfrak{g}^*$.

(1) Show that the corresponding Casimir element C defined by Proposition 6.15 can be written in the form

$$C = \sum_{\alpha \in R_+} (e_\alpha f_\alpha + f_\alpha e_\alpha) + \sum_i h_i^2,$$

where e_α, f_α are defined as in Lemma 6.42, and h_i is an orthonormal basis in $\mathfrak{h}$ with respect to $(\,,\,)$.

(2) Show that in any highest weight module with highest weight λ (not necessarily finite-dimensional), C acts by the constant

$$c_\lambda = (\lambda, \lambda + 2\rho).$$

(3) Using the arguments from the proof of Proposition 6.18, show if $(\,,\,) = K$ is the Killing form, then the corresponding Casimir element C_K acts by 1 in the adjoint representation.

(4) Let θ be the maximal root as defined in Example 8.26. Show that

$$K(\theta, \theta + 2\rho) = 1$$

and deduce from it that

$$K(\theta, \theta) = \frac{1}{2h^\vee}, \qquad h^\vee = 1 + \langle \rho, \theta^\vee \rangle.$$

(The number $h^\vee$ is called the *dual Coxeter number*.)

Since it is known that θ is always a long root (as defined in Corollary 7.51), this exercise shows that if we rescale the Killing form on $\mathfrak{g}$ by letting $\tilde{K} = \frac{1}{2h^\vee}K$, then the associated form on $\mathfrak{g}^*$ has the property $\tilde{K}(\alpha, \alpha) = 2h^\vee K(\alpha, \alpha) = 2$ for long roots α. This renormalization is commonly used, for example, in the theory of affine Lie algebras.

8.8. (1) Let $f(x)$, $x = (x_1, \ldots, x_n)$, be a polynomial in n variables. Show that if $f(x) = 0$ for all $x \in \mathbb{Z}_+^n$, then $f = 0$.

(2) Show that if $f_1, f_2 \in S\mathfrak{h}$ are such that $f_1(\lambda) = f_2(\lambda)$ for all $\lambda \in P_+$, then $f_1 = f_2$.

8.9. Let V_n be the irreducible $(n+1)$-dimensional representation of $\mathfrak{sl}(2, \mathbb{C})$ as in Theorem 4.59. Using results of Example 8.6, show that

$$V_n \otimes V_m \simeq \bigoplus V_k,$$

where the direct sum is over all $k \in \mathbb{Z}_+$ satisfying the *Clebsh–Gordan condition*

$$|n - m| \leq k \leq n + m$$

$$n + m - k \in 2\mathbb{Z}$$

8.10. Define a bilinear form $(,)_1$ on $\mathbb{C}[P]$ by

$$(f, g)_1 = \frac{1}{|W|} \int f \overline{g} \delta \overline{\delta},$$

where involution $\overline{}$ is defined by $\overline{e^\lambda} = e^{-\lambda}$, δ is the Weyl denominator (8.23), and $\int : \mathbb{C}[P] \to \mathbb{C}$ is defined by

$$\int e^\lambda = \begin{cases} 1, & \lambda = 0 \\ 0 & \text{otherwise.} \end{cases}$$

(1) Show that $(,)_1$ is symmetric.
(2) Using Weyl character formula, show that characters $\mathrm{ch}(L_\lambda)$, $\lambda \in P_+$, are orthonormal with respect to this form.

Overview of the literature

In this chapter we put together an overview of the literature and some suggestions for further reading. The list is divided into three sections: textbooks (books suitable for readers just learning the theory), monographs (books that provide detailed coverage but which still can be classified as "core" theory of Lie groups and Lie algebras) and "Further reading". Needless to say, this division is rather arbitrary and should not be taken too seriously.

Basic textbooks

There is a large number of textbooks on Lie groups and Lie algebras. Below we list some standard references which can be used either to complement the current book or to replace it.

Basic theory of Lie groups (subgroups, exponential map, etc.) can be found in any good book on differential geometry, such as Spivak [49] or Warner [55]. For more complete coverage, including discussion of representation theory, the classic references are Bröcker and tom Dieck [4] or the book by Fulton and Harris [11]. Other notable books in this category include Varadarajan [51], Onishchik and Vinberg [41]. The latest (and highly recommended) additions to this list are Bump [5], Sepanski [44] and Procesi [43]. Each of these books has its own strengths and weaknesses; we suggest that the reader looks at them to choose the book which best matches his tastes.

For Lie algebras and in particular semisimple Lie algebras, probably the best reference is Serre [47]. Without this book, the author of the current book would have never chosen Lie theory as his field of study. Another notable book about Lie algebras is Humphreys [22] (largely inspired by Serre's book).

Monographs

For readers who have learned the basic theory covered in this book or in the textbooks listed above and want to go deeper, there is no shortage of excellent in-depth books. Here are some notable titles.

For the foundations of the theory of Lie groups, the reader may consult Serre [46] and Duistermaat and Kolk [10], or the classical book by Chevalley [7]. A detailed exposition of the structure theory of Lie groups, including semisimple Lie groups, can be found in Knapp [32] or in Zhelobenko [57]; Helgason [18], in addition to providing an introduction to theory of Lie groups and Lie algebras, also includes a wealth of information about structure theory of Lie groups and homogeneous spaces.

An overview of representation theory, including the theory of infinite-dimensional representations, can be found in Kirillov [29].

Closely related to the theory of Lie groups is the theory of algebraic groups; good introduction can be found in Springer [50].

For Lie algebras, Jacobson [24] provides a comprehensive monograph on Lie algebras; in particular, there the reader can find the proofs of all the results on Lie algebras whose proof we chose to skip in our book. An equally comprehensive exposition can be found in Bourbaki [3]. For the study of universal enveloping algebras, the best source is Dixmier [9].

A detailed exposition of the theory of root systems, Weyl groups and closely related Coxeter groups can be found in Humphreys [23].

Further reading

In this section, we list some more advanced topics which might be of interest to readers who have mastered basic theory of Lie groups and Lie algebras. This list is highly biased and reflects the author's preferences; doubtless other people would suggest other topics.

Infinite-dimensional Lie groups and algebras

So far we have only discussed finite-dimensional Lie groups and Lie algebras. In general, the study of infinite-dimensional Lie groups and Lie algebras is hopelessly difficult. However, it turns out that there is a large class of infinite-dimensional Lie algebras, called Kac–Moody algebras, which in many ways are similar to semisimple finite-dimensional Lie algebras. These algebras have proved to be of vital importance to mathematical physics: they serve as the groups of symmetries in conformal field theory. A detailed exposition of these topics can be found in Kac's book [26] and those of Pressley and Segal [42] and Kumar [33].

Quantum groups

One of the most interesting developments in the theory of Lie groups and Lie algebras in recent years is related to objects which are not actually Lie algebras or groups but rather certain deformations of them. These deformations, called "quantum groups", are associative algebras where multiplication depends on an extra parameter q and which for $q = 1$ coincide with the usual universal enveloping algebra $U\mathfrak{g}$. It turns out that these quantum groups have a very interesting representation theory, with many features that do not appear for the usual Lie algebras. They also appear in many applications: to physics (where they again appear as groups of symmetries in conformal field theory), to topology (they can be used to construct invariants of knots and three-manifolds, such as the famous Jones polynomial), to combinatorics (special polynomials), and much more. A good introduction to quantum groups can be found in the books of Jantzen [25] or Kassel [28].

Analysis on homogeneous spaces

We have briefly discussed the analysis on compact Lie groups in Section 4.7. In particular, we mentioned that the Peter–Weyl theorem should be regarded as a non-commutative analog of the Fourier series.

However, this is just the beginning. One can also study various classes of functions on non-compact Lie groups, or on various homogeneous spaces for G, study invariant differential operators on such spaces, integral transforms, and much more. This is commonly referred to as "harmonic analysis on homogeneous spaces". The classical reference for the geometry of homogeneous spaces is Helgason [18]; analysis on such spaces is discussed in Helgason [20] and [19]. Other notable references include Molchanov [38] and Warner [54].

Unitary infinite-dimensional representations

So far, we mostly considered the theory of finite-dimensional representations. Again, the theory of infinite-dimensional representations in full generality is hopelessly complicated, even for semisimple Lie algebras and groups. However, the theory of unitary infinite-dimensional representations seems to be more manageable but by no means trivial. A large program of study of infinite-dimensional unitary representations of real reductive groups has been initiated by Vogan; an overview of results can be found in [53].

Special functions and combinatorics

Representation theory of Lie groups and Lie algebras is intimately related with combinatorics. For many groups, matrix coefficients and the characters of certain representations can be explicitly written in terms of classical special functions and orthogonal polynomials; thus, various results from representation theory (such as orthogonality relation for matrix coefficients) become identities involving such functions.

Representation theory of $\mathfrak{sl}(n, \mathbb{C})$ is especially closely related to combinatorics: as was mentioned in Section 8.7, irreducible representations of $\mathfrak{sl}(n, \mathbb{C})$ are parametrized by Young diagrams, which are one of the central objects of study in combinatorics, and characters of irreducible representations are Schur polynomials.

A detailed study of various links between the theory of special functions, combinatorics, and representation theory can be found in Klimyk and Vilenkin [31].

Geometric representation theory

An extremely fruitful approach to representation theory comes from geometry: instead of describing representations algebraically, by generators and relations, they are constructed in geometric terms – for example, as spaces of global sections of certain vector bundles on a manifold with the action of the group. This approach leads to some truly remarkable results. The simplest example of such a construction is the Borel–Weil theorem, which states that any irreducible finite-dimensional representation of a semisimple complex group can be obtained as a space of global sections of a certain line bundle L_λ over the corresponding flag variety; in fact, line bundles over the flag variety are classified by integral weights (see [44]).

This result has a far-reaching generalization: one can construct all highest weight modules (possibly infinite-dimensional) if we replace line bundles by appropriate generalizations – D–modules or perverse sheaves. This provides a connection between the algebraic structure of the category of highest weight modules (to be precise, category $\mathcal{O}$) and geometric structure of the cell decomposition of the flag variety; in particular, certain multiplicities in the category of highest-weight modules are equal to dimensions of the cohomology spaces of so called "Schubert cells" in the flag variety (since these cells are not manifolds but have singularities, appropriate cohomology theory is not the usual de Rham or singular cohomology, but more complicated one, called *intersection cohomology*). An introduction to this theory can be found in Miličić [37].

Another good reference for geometric methods in representation theory is Chriss and Ginzburg [8]; however, this book is more concerned with representations of Hecke algebras than Lie groups.

Appendix A

Root systems and simple Lie algebras

In this appendix, for each of the Dynkin diagrams of types $A_n, \ldots D_n$, we give an explicit description of the corresponding root system and simple Lie algebra, along with some relevant information such as the description of the Weyl group. This section contains no proofs; we refer the reader to [3], [24] for proofs and descriptions of exceptional root systems $E_6, \ldots, G_2$.

In this appendix, we use the following notation.

$\mathfrak{g}$: a complex simple Lie algebra, with fixed Cartan subalgebra $\mathfrak{h} \subset \mathfrak{g}$.

$R \subset \mathfrak{h}^*$: the root system of $\mathfrak{g}$.

$E = \mathfrak{h}^*_{\mathbb{R}}$: the real vector space spanned by roots.

$(\,,\,)$: the symmetric invariant bilinear form on $\mathfrak{h}^*$ normalized so that $(\alpha, \alpha) = 2$ for long roots.

R_+: set of positive roots (of course, there are many ways to choose it; we will only give the most common and widely used choice).

$\Pi = \{\alpha_1, \ldots, \alpha_r\}$, $r = \mathrm{rank}(R)$: set of simple roots (see Definition 7.12).

W: the Weyl group (see Section 7.2).

$P \subset E$: the weight lattice (see Section 7.5).

$Q \subset E$: the root lattice (see Section 7.5).

θ: the highest root (see Example 8.26).

$\rho = \frac{1}{2}\sum_{R_+} \alpha$ (see (7.22)).

$h = \mathrm{ht}(\theta) + 1$, $h^{\vee} = \langle \rho, \theta^{\vee} \rangle + 1$: Coxeter number and dual Coxeter number, see Exercise 8.6, Exercise 8.7.

A.1. $A_n = \mathfrak{sl}(n+1, \mathbb{C})$, $n \geq 1$

Lie algebra: $\mathfrak{g} = \mathfrak{sl}(n+1, \mathbb{C})$, with Cartan subalgebra $\mathfrak{h} = \{$diagonal matrices with trace $0\}$. We denote by $e_i \in \mathfrak{h}^*$ the functional

202

defined by

$$e_i: \begin{bmatrix} h_1 & 0 \ldots & 0 \\ & \ddots & \\ 0 & \ldots & h_{n+1} \end{bmatrix} \mapsto h_i$$

Then $\mathfrak{h}^* = \bigoplus \mathbb{C}e_i / \mathbb{C}(e_1 + \cdots + e_{n+1})$, and

$$E = \mathfrak{h}^*_{\mathbb{R}} = \bigoplus \mathbb{R}e_i / \mathbb{R}(e_1 + \cdots + e_{n+1})$$

with the inner product defined by $(\lambda, \mu) = \sum \lambda_i \mu_i$ if representatives λ, μ are chosen so that $\sum \lambda_i = \sum \mu_i = 0$.

Root system: $R = \{e_i - e_j \mid i \neq j\}$

Root subspace corresponding to root $\alpha = e_i - e_j$ is $\mathfrak{g}_\alpha = \mathbb{C}E_{ij}$, and the corresponding coroot $h_\alpha = \alpha^\vee \in \mathfrak{h}$ is $h_\alpha = E_{ii} - E_{jj}$.

Positive and simple roots: $R_+ = \{e_i - e_j \mid i < j\}$, $|R_+| = n(n+1)/2$

$\Pi = \{\alpha_1, \ldots, \alpha_n\}$, $\alpha_i = e_i - e_{i+1}$.

Dynkin diagram: O——O———————— ——O——O

Cartan matrix:

$$A = \begin{bmatrix} 2 & -1 & & & & \\ -1 & 2 & -1 & & & \\ & -1 & 2 & -1 & & \\ & & \ddots & \ddots & \ddots & \\ & & & -1 & 2 & -1 \\ & & & & -1 & 2 \end{bmatrix}$$

Weyl group: $W = S_{n+1}$, acting on E by permutations. Simple reflections are $s_i = (i\ i+1)$.

Weight and root lattices:

$P = \{(\lambda_1, \ldots, \lambda_{n+1}) \mid \lambda_i - \lambda_j \in \mathbb{Z}\}/\mathbb{R}(1, \ldots, 1) =$
$= \{(\lambda_1, \ldots, \lambda_n, 0) \mid \lambda_i \in \mathbb{Z}\}$.
$Q = \{(\lambda_1, \ldots, \lambda_{n+1}) \mid \lambda_i \in \mathbb{Z}, \sum \lambda_i = 0\}$
$P/Q \simeq \mathbb{Z}/(n+1)\mathbb{Z}$

Dominant weights and positive Weyl chamber:

$C_+ = \{\lambda_1, \ldots, \lambda_{n+1}) \mid \lambda_1 > \lambda_2 > \cdots > \lambda_{n+1}\}/\mathbb{R}(1, \ldots, 1)$
$= \{\lambda_1, \ldots, \lambda_n, 0) \mid \lambda_1 > \lambda_2 > \cdots > \lambda_n > 0\}$.

$$P_+ = \{(\lambda_1, \ldots, \lambda_{n+1}) \mid \lambda_i - \lambda_{i+1} \in \mathbb{Z}_+\}/\mathbb{R}(1, \ldots, 1)$$
$$= \{(\lambda_1, \ldots, \lambda_n, 0) \mid \lambda_i \in \mathbb{Z}, \lambda_1 \geq \lambda_2 \cdots \geq \lambda_n \geq 0\}.$$

Maximal root, ρ, and the Coxeter number:
$$\theta = e_1 - e_{n+1} = (1, 0, \ldots, 0, -1)$$
$$\rho = (n, n-1, \ldots, 1, 0) = (n/2, (n-2)/2, \ldots, (-n)/2)$$
$$h = h^\vee = n+1$$

A.2. $B_n = \mathfrak{so}(2n+1, \mathbb{C})$, $n \geq 1$

Lie algebra:

$\mathfrak{g} = \mathfrak{so}(2n+1, \mathbb{C})$, with Cartan subalgebra consisiting of block-diagonal matrices

$$\mathfrak{h} = \left\{ \begin{bmatrix} A_1 & & & & \\ & A_2 & & & \\ & & \ddots & & \\ & & & A_n & \\ & & & & 0 \end{bmatrix} \right\}, \quad A_i = \begin{bmatrix} 0 & a_i \\ -a_i & 0 \end{bmatrix}$$

Lie algebra (alternative description):

$\mathfrak{g} = \mathfrak{so}(B) = \{a \in \mathfrak{gl}(2n+1, \mathbb{C}) \mid a + B^{-1}\,{}^t a B = 0\}$, where B is the symmetric non-degenerate bilinear form on $\mathbb{C}^{2n+1}$ with the matrix

$$B = \begin{bmatrix} 0 & I_n & 0 \\ I_n & 0 & 0 \\ 0 & 0 & 1 \end{bmatrix}$$

This Lie algebra is isomorphic to the usual $\mathfrak{so}(2n+1, \mathbb{C})$; the isomorphism is given by $a \mapsto Ba$.

In this description, the Cartan subalgebra is

$$\mathfrak{h} = \mathfrak{g} \cap \{\text{diagonal matrices}\} = \{\mathrm{diag}(x_1, \ldots, x_n, -x_1, \ldots, -x_n, 0)\}$$

Define $e_i \in \mathfrak{h}^*$ by

$$e_i \colon \mathrm{diag}(x_1, \ldots, x_n, -x_1, \ldots, -x_n, 0) \mapsto x_i.$$

Then e_i, $i = 1 \ldots n$, form a basis in $\mathfrak{h}^*$. The bilinear form is defined by $(e_i, e_j) = \delta_{ij}$.

Root system:

$R = \{\pm e_i \pm e_j \ (i \neq j), \pm e_i\}$ (signs are chosen independently)

The corresponding root subspaces and coroots in $\mathfrak{g}$ (using the alternative description) are given by

- For $\alpha = e_i - e_j$: $\mathfrak{g}_\alpha = \mathbb{C}(E_{ij} - E_{j+n,i+n})$, $h_\alpha = H_i - H_j$.
- For $\alpha = e_i + e_j$: $\mathfrak{g}_\alpha = \mathbb{C}(E_{i,j+n} - E_{j,i+n})$, $h_\alpha = H_i + H_j$.
- For $\alpha = -e_i - e_j$: $\mathfrak{g}_\alpha = \mathbb{C}(E_{i+n,j} - E_{j+n,i})$, $h_\alpha = -H_i - H_j$.
- For $\alpha = e_i$, $\mathfrak{g}_\alpha = \mathbb{C}(E_{i,2n+1} - E_{2n+1,n+i})$, $h_\alpha = 2H_i$.
- For $\alpha = -e_i$, $\mathfrak{g}_\alpha = \mathbb{C}(E_{n+i,2n+1} - E_{2n+1,i})$, $h_\alpha = -2H_i$

where $H_i = E_{ii} - E_{i+n,i+n}$.

Positive and simple roots: $R_+ = \{e_i \pm e_j \ (i < j), e_i\}$, $|R_+| = n^2$

$\Pi = \{\alpha_1, \ldots, \alpha_n\}$, $\alpha_1 = e_1 - e_2, \ldots, \alpha_{n-1} = e_{n-1} - e_n$, $\alpha_n = e_n$.

Dynkin diagram:

Cartan matrix:

$$
A = \begin{bmatrix}
2 & -1 & 0 & & & & \\
-1 & 2 & -1 & & & & \\
 & -1 & 2 & -1 & & & \\
 & & & \ddots & \ddots & \ddots & \\
 & & & & -1 & 2 & -1 \\
 & & & & & -2 & 2
\end{bmatrix}
$$

Weyl group: $W = S_n \ltimes (\mathbb{Z}_2)^n$, acting on E by permutations and sign changes of coordinates. Simple reflections are $s_i = (i \ i+1)$ $(i = 1 \ldots n-1)$, $s_n : (\lambda_1, \ldots, \lambda_n) \mapsto (\lambda_1, \ldots, -\lambda_n)$.

Weight and root lattices: (in basis $e_i \ldots, e_n$)

$P = \{(\lambda_1, \ldots, \lambda_n) \mid \lambda_i \in \frac{1}{2}\mathbb{Z}, \lambda_i - \lambda_j \in \mathbb{Z}\}$

$Q = \mathbb{Z}^n$

$P/Q \simeq \mathbb{Z}_2$

Dominant weights and positive Weyl chamber:

$C_+ = \{\lambda_1, \ldots, \lambda_n) \mid \lambda_1 > \lambda_2 > \cdots > \lambda_n > 0\}$.

$P_+ = \{(\lambda_1, \ldots, \lambda_n) \mid \lambda_1 \geq \lambda_2 \geq \cdots \geq \lambda_n \geq 0, \lambda_i \in \frac{1}{2}\mathbb{Z}, \lambda_i - \lambda_j \in \mathbb{Z}\}$.

Maximal root, ρ, and the Coxeter number:

$\theta = e_1 + e_2 = (1, 1, 0, \ldots, 0)$

$\rho = (n - \frac{1}{2}, n - \frac{3}{2}, \ldots, \frac{1}{2})$

$h = 2n$, $h^\vee = 2n - 1$ (for $n \geq 2$; for $n = 1$, $\mathfrak{so}(3, \mathbb{C}) \simeq \mathfrak{sl}(2, \mathbb{C})$, so $h = h^\vee = 2$).

A.3. $C_n = \mathfrak{sp}(n, \mathbb{C}), n \geq 1$

Lie algebra: $\mathfrak{g} = \mathfrak{sp}(n, \mathbb{C}) = \{a \in \mathfrak{gl}(2n, \mathbb{C}) \mid a + J^{-1}a^t J = 0\}$, where J is the skew-symmetric nondegenerate matrix

$$J = \begin{bmatrix} 0 & I_n \\ -I_n & 0 \end{bmatrix}$$

The Cartan subalgebra is given by

$$\mathfrak{h} = \mathfrak{g} \cap \{\text{diagonal matrices}\} = \{\text{diag}(x_1, \ldots, x_n, -x_1, \ldots, -x_n)\}$$

Define $e_i \in \mathfrak{h}^*$ by

$$e_i: \text{diag}(x_1, \ldots, x_n, -x_1, \ldots, -x_n) \mapsto x_i.$$

Then $e_i, i = 1 \ldots n$, form a basis in $\mathfrak{h}^*$. The bilinear form is defined by $(e_i, e_j) = \frac{1}{2}\delta_{ij}$.

Root system:

$R = \{\pm e_i \pm e_j \ (i \neq j), \pm 2e_i\}$ (signs are chosen independently)

The corresponding root subspaces and coroots are given by

- For $\alpha = e_i - e_j$: $\mathfrak{g}_\alpha = \mathbb{C}(E_{ij} - E_{j+n,i+n})$, $h_\alpha = H_i - H_j$.
- For $\alpha = e_i + e_j$: $\mathfrak{g}_\alpha = \mathbb{C}(E_{i,j+n} + E_{j,i+n})$, $h_\alpha = H_i + H_j$.
- For $\alpha = -e_i - e_j$: $\mathfrak{g}_\alpha = \mathbb{C}(E_{i+n,j} + E_{j+n,i})$, $h_\alpha = -H_i - H_j$.
- For $\alpha = 2e_i$, $\mathfrak{g}_\alpha = \mathbb{C}E_{i,i+n}$, $h_\alpha = H_i$
- For $\alpha = -2e_i$, $\mathfrak{g}_\alpha = \mathbb{C}E_{i+n,i}$, $h_\alpha = -H_i$

where $H_i = E_{ii} - E_{i+n,i+n}$.

Positive and simple roots: $R_+ = \{e_i \pm e_j \ (i < j), 2e_i\}$, $|R_+| = n^2$

$\Pi = \{\alpha_1, \ldots, \alpha_n\}$, $\alpha_1 = e_1 - e_2, \ldots, \alpha_{n-1} = e_{n-1} - e_n, \alpha_n = 2e_n$.

Dynkin diagram: ○———○— $\cdots$ —○———○⇐○

Cartan matrix:

$$A = \begin{bmatrix} 2 & -1 & & & & \\ -1 & 2 & -1 & & & \\ & -1 & 2 & -1 & & \\ & & \ddots & \ddots & \ddots & \\ & & & -1 & 2 & -2 \\ & & & & -1 & 2 \end{bmatrix}$$

Weyl group: $W = S_n \ltimes (\mathbb{Z}_2)^n$, acting on E by permutations and sign changes of coordinates. Simple reflections are $s_i = (i\ i+1)$ $(i = 1 \ldots n-1)$, $s_n : (\lambda_1, \ldots, \lambda_n) \mapsto (\lambda_1, \ldots, -\lambda_n)$.

Weight and root lattices: (in basis $e_1, \ldots, e_n$)

$P = \mathbb{Z}^n$

$Q = \{(\lambda_1, \ldots, \lambda_n) \mid \lambda_i \in \mathbb{Z}, \sum \lambda_i \in 2\mathbb{Z}\}$

$P/Q \simeq \mathbb{Z}_2$

Dominant weights and positive Weyl chamber:

$C_+ = \{\lambda_1, \ldots, \lambda_n) \mid \lambda_1 > \lambda_2 > \cdots > \lambda_n > 0\}.$

$P_+ = \{(\lambda_1, \ldots, \lambda_n) \mid \lambda_1 \geq \lambda_2 \geq \cdots \geq \lambda_n \geq 0, \lambda_i \in \mathbb{Z}\}.$

Maximal root, ρ, and the Coxeter number:

$\theta = 2e_1 = (2, 0, \ldots, 0)$

$\rho = (n, n-1, \ldots, 1)$

$h = 2n, h^\vee = n + 1$

A.4. $D_n = \mathfrak{so}(2n, \mathbb{C})$, $n \geq 2$

Lie algebra: $\mathfrak{g} = \mathfrak{so}(2n, \mathbb{C})$, with Cartan subalgebra consisting of block-diagonal matrices

$$\mathfrak{h} = \left\{ \begin{bmatrix} A_1 & & & \\ & A_2 & & \\ & & \ddots & \\ & & & A_n \end{bmatrix} \right\}, \qquad A_i = \begin{bmatrix} 0 & h_i \\ -h_i & 0 \end{bmatrix}$$

Lie algebra (alternative description):

$\mathfrak{g} = \mathfrak{so}(B) = \{a \in \mathfrak{gl}(2n, \mathbb{C}) \mid a + B^{-1}a^t B = 0\}$, where B is the symmetric non-degenerate bilinear form on $\mathbb{C}^{2n}$ with the matrix

$$B = \begin{bmatrix} 0 & I_n \\ I_n & 0 \end{bmatrix}$$

This Lie algebra is isomorphic to the usual $\mathfrak{so}(2n, \mathbb{C})$; the isomorphism is given by $a \mapsto Ba$.

In this description, the Cartan subalgebra is

$$\mathfrak{h} = \mathfrak{g} \cap \{\text{diagonal matrices}\} = \{\mathrm{diag}(x_1, \ldots, x_n, -x_1, \ldots, -x_n)\}$$

Define $e_i \in \mathfrak{h}^*$ by

$$e_i\colon \ \mathrm{diag}(x_1,\ldots,x_n,-x_1,\ldots,-x_n) \mapsto x_i.$$

Then $e_i, i = 1\ldots n$ form a basis in $\mathfrak{h}^*$. The bilinear form is given by $(e_i, e_j) = \delta_{ij}$.

Root system:

$R = \{\pm e_i \pm e_j \ (i \neq j)\}$ (signs are chosen independently)

The corresponding root subspaces and coroots in $\mathfrak{g}$ (using the alternative description) are given by

- For $\alpha = e_i - e_j$: $\mathfrak{g}_\alpha = \mathbb{C}(E_{ij} - E_{j+n,i+n})$, $h_\alpha = H_i - H_j$.
- For $\alpha = e_i + e_j$: $\mathfrak{g}_\alpha = \mathbb{C}(E_{i,j+n} - E_{j,i+n})$, $h_\alpha = H_i + H_j$.
- For $\alpha = -e_i - e_j$: $\mathfrak{g}_\alpha = \mathbb{C}(E_{i+n,j} - E_{j+n,i})$, $h_\alpha = -H_i - H_j$

where $H_i = E_{ii} - E_{i+n,i+n}$.

Positive and simple roots: $R_+ = \{e_i \pm e_j \ (i < j)\}$, $|R_+| = n(n-1)$

$\Pi = \{\alpha_1,\ldots,\alpha_n\}$, $\alpha_1 = e_1 - e_2,\ldots,\alpha_{n-1} = e_{n-1} - e_n$, $\alpha_n = e_{n-1} + e_n$.

Dynkin diagram:

Cartan matrix:

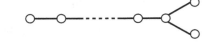

$$A = \begin{bmatrix} 2 & -1 & & & & & \\ -1 & 2 & -1 & & & & \\ & -1 & 2 & -1 & & & \\ & & & \ddots & \ddots & \ddots & \\ & & & -1 & 2 & -1 & -1 \\ & & & & -1 & 2 & \\ & & & & -1 & & 2 \end{bmatrix}$$

Weyl group: $W = \{$permutations and even number of sign changes$\}$. Simple reflections are $s_i = (i \ i+1)$, $i = 1\ldots n-1$, $s_n\colon (\lambda_1,\ldots,\lambda_{n-1},\lambda_n) \mapsto (\lambda_1,\ldots,-\lambda_n,-\lambda_{n-1})$.

Weight and root lattices: (in basis $e_1,\ldots,e_n$)

$P = \{(\lambda_1,\ldots,\lambda_n) \mid \lambda_i \in \frac{1}{2}\mathbb{Z}, \lambda_i - \lambda_j \in \mathbb{Z}\}$

$Q = \{(\lambda_1,\ldots,\lambda_n) \mid \lambda_i \in \mathbb{Z}, \sum \lambda_i \in 2\mathbb{Z}\}$

$P/Q \simeq \mathbb{Z}_2 \times \mathbb{Z}_2$ for even n, $P/Q \simeq \mathbb{Z}_4$ for odd n

Dominant weights and positive Weyl chamber:

$C_+ = \{(\lambda_1,\ldots,\lambda_n) \mid \lambda_1 > \lambda_2 > \cdots > \lambda_n, \lambda_{n-1} + \lambda_n > 0\}$.

$P_+ = \{(\lambda_1,\ldots,\lambda_n) \mid \lambda_1 \geq \lambda_2 \geq \ldots$

$$\geq \lambda_n, \lambda_{n-1} + \lambda_n \geq 0,$$
$$\lambda_i \in \tfrac{1}{2}\mathbb{Z}, \lambda_i - \lambda_j \in \mathbb{Z}\}.$$

Maximal root, ρ, and the Coxeter number:

$$\theta = e_1 + e_2 = (1, 1, 0, \ldots, 0)$$
$$\rho = (n - 1, n - 2, \ldots, 0)$$
$$h = h^\vee = 2n - 2$$

Appendix B

Sample syllabus

In this section, we give a sample syllabus of a one-semester graduate course on Lie groups and Lie algebras based on this book. This course is designed to fit the standard schedule of US universities: 14 week semester, with two lectures a week, each lecture 1 hour and 20 minutes long.

Lecture 1: Introduction. Definition of a Lie group; C^1 implies analytic. Examples: $\mathbb{R}^n$, S^1, SU(2). Theorem about closed subgroup (no proof). Connected component and universal cover.

Lecture 2: G/H. Action of G on manifolds; homogeneous spaces. Action on functions, vector fields, etc. Left, right, and adjoint action. Left, right, and bi-invariant vector fields (forms, etc).

Lecture 3: Classical groups: GL, SL, SU, SO, Sp – definition. Exponential and logarithmic maps for matrix groups. Proof that classical groups are smooth; calculation of the corresponding Lie algebra and dimension. Topological information (connectedness, π_1). One-parameter subgroups in a Lie group: existence and uniqueness.

Lecture 4: Lie algebra of a Lie groups:

$$\mathfrak{g} = T_1 G = \text{right-invariant vector fields} = \text{1-parameter subgroups}.$$

Exponential and logarithmic maps and their properties. Morphisms $f : G_1 \rightarrow G_2$ are determined by $f_* : \mathfrak{g}_1 \rightarrow \mathfrak{g}_2$. Example: elements $J_x, J_y, J_z \in \mathfrak{so}(3)$. Definition of commutator: $e^x e^y = e^{x+y+\frac{1}{2}[x,y]+\cdots}$.

Lecture 5: Properties of the commutator. Relation with the group commutator; Ad and ad. Jacobi identity. Abstract Lie algebras and morphisms.

$[x, y] = xy - yx$ for matrix algebras. Relation with the commutator of vector fields. Campbell–Hausdorff formula (no proof).

Lecture 6: If G_1 is simply-connected, then $\mathrm{Hom}(G_1, G_2) = \mathrm{Hom}(\mathfrak{g}_1, \mathfrak{g}_2)$. Analytic subgroups and Lie subalgebras. Ideals in $\mathfrak{g}$ and normal subgroups in G.

Lecture 7: Lie's third theorem (no proof). Corollary: category of connected, simply-connected Lie groups is equivalent to the category of Lie algebras. Representations of G = representations of $\mathfrak{g}$. Action by vector fields.

Example: representations of $\mathrm{SO}(3), \mathrm{SU}(2)$. Complexification; $\mathfrak{su}(n)$ and $\mathfrak{sl}(n)$.

Lecture 8: Representations of Lie groups and Lie algebras. Subrepresentations, direct sums, $V_1 \otimes V_2$, V^*, action on $\mathrm{End}\, V$. Irreducibility. Intertwining operators. Schur lemma. Semisimplicity.

Lecture 9: Unitary representations. Complete reducibility of representation for a group with invariant integral. Invariant integral for finite groups and for compact Lie groups; Haar measure. Example: representations of S^1 and Fourier series.

Lecture 10: Characters and Peter–Weyl theorem.

Lecture 11: Universal enveloping algebra. Central element $J_x^2 + J_y^2 + J_z^2 \in U\,\mathfrak{so}(3, \mathbb{R})$. Statement of PBW theorem.

Lecture 12: Structure theory of Lie algebras: generalities. Commutant. Solvable and nilpotent Lie algebras: equivalent definitions. Example: upper triangular matrices. Lie theorem (about representations of a solvable Lie algebra).

Lecture 13: Engel's theorem (without proof). Radical. Semisimple Lie algebras. Example: semisimplicity of $\mathfrak{sl}(2)$. Levi theorem (without proof). Statement of Cartan criterion of solvability and semisimplicity.

Lecture 14: Jordan decomposition (into semisimple and nilpotent element). Proof of Cartan criterion.

Lecture 15: Corollaries: every semisimple algebra is direct sum of simple ones; ideal, quotient of a semisimple algebra is semisimple; $[\mathfrak{g}, \mathfrak{g}] = \mathfrak{g}$; every derivation is inner.

Relation between semisimple Lie algebras and compact groups.

Lecture 16: Complete reducibility of representations of a semisimple Lie algebra.

Lecture 17: Representations of $\mathfrak{sl}(2, \mathbb{C})$. Semisimple elements in a Lie algebra.

Lecture 18: Semisimple and nilpotent elements; Jordan decomposition. Toral subalgebras. Definition of Cartan (maximal toral) subalgebra. Theorem: conjugacy of Cartan subalgebras (no proof).

Lecture 19: Root decomposition and root system for semisimple Lie algebra. Basic properties. Example: $\mathfrak{sl}(n, \mathbb{C})$.

Lecture 20: Definition of an abstract root system. Weyl group. Classification of rank 2 root systems.

Lecture 21: Positive roots and simple roots. Polarizations and Weyl chambers. Transitivity of action of W on the set of Weyl chambers.

Lecture 22: Simple reflections. Reconstructing root system from set of simple roots. Length $l(w)$ and its geometric interpretation as number of separating hyperplanes.

Lecture 23: Cartan matrix and Dynkin diagrams. Classification of Dynkin diagrams (partial proof).

Lecture 24: Constructing a semisimple Lie algebra from a root system. Serre relations and Serre theorem (no proof). Classification of simple Lie algebras.

Lecture 25: Finite-dimensional representations of a semi-simple Lie algebra. Weights; symmetry under Weyl group. Example: $\mathfrak{sl}(3, \mathbb{C})$. Singular vectors.

Lecture 26: Verma modules and irreducible highest weight modules. Dominant weights and classification of finite-dimensional highest weight modules (without proof)

Lecture 27: BGG resolution and Weyl character formula

Lecture 28: Example: representations of $\mathfrak{sl}(n, \mathbb{C})$.

List of notation

$\mathbb{R}$: real numbers

$\mathbb{C}$: complex numbers

$\mathbb{K}$: either $\mathbb{R}$ or $\mathbb{C}$. This notation is used when a result holds for both $\mathbb{R}$ and $\mathbb{C}$.

$\mathbb{Z}$: integer numbers

$\mathbb{Z}_+ = \{0, 1, 2, \dots\}$: non-negative integer numbers

Linear algebra

V^*: dual vector space

$\langle\,,\,\rangle\colon V \otimes V^* \to \mathbb{C}$: canonical pairing of V with V^*.

$\mathrm{Hom}(V, W)$: space of linear maps $V \to W$

$\mathrm{End}(V) = \mathrm{Hom}(V, V)$: space of linear maps $V \to V$ considered as an associative algebra

$\mathfrak{gl}(V) = \mathrm{Hom}(V, V)$: space of linear maps $V \to V$ considered as a Lie algebra, see Example 3.14

$\mathrm{tr}\, A$: trace of a linear operator

$\mathrm{Ker}\, B = \{v \mid B(v, w) = 0 \text{ for all } w\}$, for a symmetric bilinear form B: kernel, or radical, of B

A^t: adjoint operator: if $A\colon V \to W$ is a linear operator, then $A^t\colon W^* \to V^*$.

$A = A_s + A_n$: Jordan decomposition of an operator A, see Theorem 5.59

Differential geometry

$T_p M$: tangent space to manifold M at point p

$\mathrm{Vect}(M)$: space of C^∞ vector fields on M

$\mathrm{Diff}(M)$: group of diffeomorphisms of a manifold M

Φ_ξ^t: time t flow of vector field ξ, see Section 3.5

Lie groups and Lie algebras

G_m: stabilizer of point m, see (2.3)

$\mathfrak{g} = \text{Lie}(G)$: Lie algebra of group G, see Theorem 3.20

$\exp\colon \mathfrak{g} \to G$: exponential map, see Definition 3.2

$\text{ad}\, x.y = [x, y]$, see (2.4)

$\mathfrak{z}(\mathfrak{g})$: center of $\mathfrak{g}$, see Definition 3.34

$\text{Der}(\mathfrak{g})$: Lie algebra of derivations of $\mathfrak{g}$, see (3.14)

$[\mathfrak{g}, \mathfrak{g}]$: commutant of $\mathfrak{g}$, see Definition 5.19

$\text{rad}(\mathfrak{g})$: radical of Lie algebra $\mathfrak{g}$, see Proposition 5.39

$K(x, y)$: Killing form, see Definition 5.50

$\text{Ad}\, g$: adjoint action of G on $\mathfrak{g}$, see (2.4)

$U\mathfrak{g}$: universal enveloping algebra, see Definition 5.1

Representations

$\text{Hom}_G(V, W)$, $\text{Hom}_{\mathfrak{g}}(V, W)$: spaces of intertwining operators, see Definition 4.1

χ_V: character of representation V, see Definition 4.43

V^G, $V^{\mathfrak{g}}$: spaces of invariants, see Definition 4.13

Semisimple Lie algebras and root systems

$\mathfrak{h}$: Cartan subalgebra, see Definition 6.32

$\mathfrak{g}_\alpha$: root subspace, see Theorem 6.38

$R \subset \mathfrak{h}^* \setminus \{0\}$: root system

$h_\alpha = \alpha^\vee = 2H_\alpha/(\alpha, \alpha) \in \mathfrak{h}$: dual root, see (6.5), (6.4) (for root system of a Lie algebra) and (7.4) for an abstract root system

$\text{rank}(\mathfrak{g}) = \dim \mathfrak{h}$: rank of a semisimple Lie algebra, see (6.1)

s_α: reflection defined by a root α, see Definition 7.1

$R_\pm$: positive and negative roots, see (7.6)

$\Pi = \{\alpha_1, \ldots, \alpha_r\} \subset R_+$: simple roots, see Definition 7.12

$\text{ht}(\alpha)$: height of a positive root, see (7.8)

$L_\alpha = \{\lambda \in E \mid (\lambda, \alpha) = 0\}$: root hyperplane, see (7.15)

C_+: positive Weyl chamber, see (7.17)

W: Weyl group, see Definition 7.6

$s_i = s_{\alpha_i}$: simple reflection, see Theorem 7.30

$l(w)$: length of element $w \in W$, see Definition 7.34, Theorem 7.37

w_0: longest element of the Weyl group, see Lemma 7.39

$\rho = \frac{1}{2} \sum_{R_+} \alpha$, see Lemma 7.36

$\mathfrak{n}_\pm$: nilpotent subalgebras, see (7.25)

P: weight lattice, see (7.12)
Q: root lattice, see (7.9)
$Q_+ = \{\sum n_i \alpha_i, n_i \in \mathbb{Z}_+\}$, see (8.10)

Representations of semisimple Lie algebras

$V[\lambda]$: weight subspace, see Definition 4.54, Definition 8.1
$\mathbb{C}[P]$: group algebra of the weight lattice, see (8.4)
M_λ: Verma module, see (8.7)
L_λ: irreducible highest weight representation, see Theorem 8.18
$\prec$: partial order on weights, see (8.11)
$w.\lambda$: shifted action of Weyl group on weights, see (8.20)
$\mathfrak{b}$: Borel subalgebra, see (8.8)

Bibliography

1. Arnold, V. *Mathematical Methods of Classical Mechanics*, second edition, Graduate Texts in Mathematics, Vol. 60. New York: Springer-Verlag, 1989.
2. Bernstein, I. N., Gelfand, I. M., Gelfand, S. I. Differential operators on the base affine space and a study of g-modules. In: *Lie Groups and their Representations, Proc. Summer School on Group Representations of the János Bolyai Math. Soc., Budapest, 1971*, pp. 21–64. New York: Halsted, 1975.
3. Bourbaki, N. *Lie Groups and Lie Algebras*, Chapters 1–3. Berlin: Springer, 1998; *Lie Groups and Lie Algebras*, Chapters 4–6. Berlin: Springer, 2002.
4. Bröcker, T., tom Dieck, T. *Representations of Compact Lie Groups*, Graduate Texts in Mathematics, Vol. 98. New York: Springer, 1985.
5. Bump, D. *Lie Groups*, Graduate Texts in Mathematics, Vol. 225. New York: Springer, 2004.
6. Carter, R., Segal, G., Macdonald, I. *Lectures on Lie Groups and Lie Algebras*, London Mathematical Society Student Texts, 32. Cambridge: Cambridge University Press, 1995.
7. Chevalley, C. *Theory of Lie Groups*. Princeton, NJ: Princeton University Press, 1946.
8. Chriss, N., Ginzburg, V. *Representation Theory and Complex Geometry*. Boston, MA: Birkhäuser Boston, Inc., 1997.
9. Dixmier, J. *Enveloping Algebras*. Providence, RI: American Mathematical Society, 1996. [This is a reprint of 1977 translation of 1974 French original.]
10. Duistermaat, J. J., Kolk, J. A. C. *Lie Groups*. Berlin: Springer-Verlag, 2000.
11. Fulton, W., Harris, J. *Representation Theory: A First Course*, Graduate Texts in Mathematics, Vol. 129. New York: Springer-Verlag, 1991.
12. Feigin, B. L., Fuchs, D. B. Cohomologies of Lie groups and Lie algebras. In: *Lie Groups and Lie Algebras II*, Encyclopaedia of Mathematical Sciences, Vol. 21. Berlin: Springer, 2000.
13. *GAP – Groups, Algorithms, Programming – a System for Computational Discrete Algebra*, available from www.gap-system.org.
14. Greub, W., Halperin, S., Vanstone, R. *Connections, Curvature, and Cohomology*. New York: Academic Press, 1976.

15. Gorbatsevich, V. V., Onischik, A. L., Vinberg, E. B. Structure of Lie groups and Lie algebras. In: *Lie Groups and Lie Algebras III*, Encyclopaedia of Mathematical Sciences, Vol. 41. Berlin: Springer, 1994.

16. Hall, B. *Lie Groups, Lie Algebras, and Representations. An Elementary Introduction*, Graduate Texts in Mathematics, Vol. 222. Berlin: Springer, 2003. Corr. 2nd printing, 2004.

17. Hatcher, A. *Algebraic Topology*. Cambridge: Cambridge University Press, 2002.

18. Helgason, S. *Differential Geometry, Lie Groups, and Symmetric Space*. New York: Academic Press, 1978.

19. Helgason, S. *Groups and Geometric Analysis*, Mathematical Surveys and Monographs, 83. Providence, RI: American Mathematical Society, 2000.

20. Helgason, S. *Geometric Analysis on Symmetric Spaces*. Mathematical Surveys and Monographs, 39. Providence, RI: American Mathematical Society, 1994.

21. Hilton, P., Stammbach, U. *A Course in Homological Algebra*, second edition, Graduate Texts in Mathematics, 4. New York–Berlin: Springer-Verlag, 1997.

22. Humphreys, J. *Introduction to Lie Algebras and Representation Theory*, second printing, revised, Graduate Texts in Mathematics, 9. New York–Berlin: Springer-Verlag, 1978.

23. Humphreys, J. *Reflection Groups and Coxeter Groups*. Cambridge: Cambridge University Press, 1990.

24. Jacobson, N. *Lie Algebras*, Republication of the 1962 original. New York: Dover Publications, Inc., 1979.

25. Jantzen, J. C. *Lectures on Quantum Groups*, Graduate Studies in Mathematics, Vol. 6. Providence, RI: American Mathematical Society, 1996.

26. Kac, V. G. *Infinite-dimensional Lie Algebras*, third edition. Cambridge: Cambridge University Press, 1990.

27. Kaplansky, I. *Lie Algebras and Locally Compact Groups*, Chicago Lectures in Mathematics. Chicago, IL: University of Chicago Press, 1995.

28. Kassel, C. *Quantum Groups*, Graduate Texts in Mathematics, 155. New York: Springer-Verlag, 1995.

29. Kirillov, A. A. *Elements of the Theory of Representations*. Berlin–New York: Springer-Verlag, 1976.

30. Kirillov, A. A. *Lectures on the Orbit Method*, Graduate Studies in Mathematics, Vol. 64. Providence, RI: American Mathematical Society, 2004.

31. Klimyk, A. U., Vilenkin, N. Ya. Representations of Lie groups and special functions. In: *Representation Theory and Noncommutative Harmonic Analysis. II. Homogeneous Spaces, Representations and Special Functions*, Encyclopaedia of Mathematical Sciences, Vol. 59. Berlin: Springer-Verlag, 1995.

32. Knapp, A. *Lie Goups Beyond an Introduction*, second edition, Progress in Mathematics, Vol. 140. Boston, MA: Birkhäuser Boston, Inc., 2002.

33. Kumar, S. *Kac-Moody Groups, their Flag Varieties and Representation Theory*, Progress in Mathematics, Vol. 204. Boston, MA: Birkhäuser Boston, Inc., 2002.

34. Landau, L. D. *Quantum Mechanics: non-Relativistic Theory. Course of Theoretical Physics, Vol. 3* third edition. London–Paris: Pergamon Press, 1981.

35. LiE: A Computer algebra package for Lie group computations, available from http://young.sp2mi.univ-poitiers.fr/~marc/LiE/

36. Macdonald, I. G. *Symmetric Functions and Hall Polynomials*, second edition. Oxford: Oxford University Press, 1995.
37. Miličić, D. Algebraic *D*-modules and representation theory of semisimple Lie groups. In: *The Penrose Transform and Analytic Cohomology in Representation Theory (South Hadley, MA, 1992)*, Contemporary Mathematics, 154, pp. 133–168. Providence, RI: American Mathematical Society, 1993.
38. Molchanov, V. E. Harmonic analysis on homogeneous spaces. In: *Representation Theory and Non-commutative Harmonic Analysis. II. Homogeneous Spaces, Representations and Special Functions*, Encyclopaedia of Mathematical Sciences, Vol. 59. Berlin: Springer-Verlag, 1995.
39. Montgomery, D., Zippin, L. *Topological Transformation Groups*, reprint of the 1955 original. Huntington, NY: Robert E. Krieger Publishing Co., 1974.
40. Mumford, D., Fogarty, J., Kirwan, F. *Geometric Invariant Theory*, third edition. Berlin: Springer-Verlag, 1994.
41. Onishchik, A. L., Vinberg, E. B. *Foundations of Lie theory*. In: *Lie Groups and Lie Algebras II*, Encyclopaedia of Mathematical Sciences, Vol. 20. Berlin: Springer-Verlag, 1990.
42. Pressley, A., Segal, G. *Loop Groups*. New York: The Clarendon Press, Oxford University Press, 1986.
43. Procesi, C. *Lie Groups: an Approach Through Invariants and Representations*. New York: Springer, 2007.
44. Sepanski, M. *Compact Lie Groups*, Graduate Texts in Mathematics, 235. New York: Springer-Verlag, 2007.
45. Serre, J.-P. *Linear Representations of Finite Groups*, Graduate Texts in Mathematics, 42. New York–Heidelberg: Springer-Verlag, 1977.
46. Serre, J.-P. *Lie Algebras and Lie Groups*, 1964 lectures given at Harvard University, second edition, Lecture Notes in Mathematics, 1500. Berlin: Springer-Verlag, 1992.
47. Serre, J.-P. *Complex Semisimple Lie Algebras*. Berlin: Springer-Verlag, 2001.
48. Simon, B. *Representations of Finite and Compact Groups*. Providence, RI: American Mathematical Society, 1996.
49. Spivak, M. *A Comprehensive Introduction to Differential Geometry*, Vol. I. second edition. Wilmington, DE: Publish or Perish, Inc., 1979.
50. Springer, T. A. *Linear Algebraic Groups*, second edition. Progress in Mathematics, 9. Boston, MA: Birkhäuser Boston, Inc., 1998.
51. Varadarajan, V. S. *Lie Groups, Lie Algebras, and their Representations*, reprint of the 1974 edition. Graduate Texts in Mathematics, 102. New York: Springer-Verlag, 1984.
52. Varadarajan, V. S. *An Introduction to Harmonic Analysis on Semisimple Lie Groups*, Cambridge Studies in Advanced Mathematics, 16. Cambridge: Cambridge University Press, 1989.
53. Vogan, Jr, D. A. *Unitary Representations of Reductive Lie Groups*, Annals of Mathematics Studies, 118. Princeton, NJ: Princeton University Press, 1987.
54. Warner, G. *Harmonic Analysis on Semi-simple Lie Groups. I, II* Die Grundlehren der mathematischen Wissenschaften, Band 188, 189. New York–Heidelberg: Springer-Verlag, 1972.

55. Warner, F. *Foundations of Differentiable Manifolds and Lie Groups*. Berlin: Springer-Verlag, 1983.
56. Weyl package for Maple, by John Stembridge. Available from `http://www.math.lsa.umich.edu/~jrs/maple.html`
57. Želobenko, D. P. *Compact Lie Groups and their Representations*, Translations of Mathematical Monographs, Vol. 40. Providence, RI: American Mathematical Society, 1973.

Index

action
    of a Lie group on a manifold 11
    left 14
    right 14
    adjoint 14, 54
    coadjoint 55
Ado theorem 42

Bruhat order 177
Borel subalgebra 168
Bernstein–Gelfand–Gelfand (BGG)
    resolution 176

Campbell–Hausdorff formula 39
Cartan's criterion
    of solvability 102
    of semisimplicity 102
Cartan subalgebra 119
Cartan matrix 151
Casimir operator 112
character 67, 165
Clebsh–Gordan condition 195
commutant 91
commutator 29
    of vector fields 34
complexification 45
coroot 133
    lattice 140
coset space 8
Coxeter relations 161
Coxeter number 194
    dual 195

derivations
    of an associative algebra 37
    of a Lie algebra 38
    inner 49
distribution 43
Dynkin diagram 151
    simply-laced 154

Engel's theorem 96
exponential map
    for matrix algebras 17
    for arbitrary Lie algebra 26

flag manifold 13
Frobenius integrability criterion 43

Haar measure 64
Harish–Chandra isomorphism 190
Heisenberg algebra 50
height 139
highest weight 72, 167
highest weight vector 72, 167
highest weight representation 167
homogeneous space 13

ideal (in a Lie algebra) 32
integrable representation 192
intertwining operator 52
invariant bilinear form 56

Jacobi identity 31